# Varn System

## Original Concept and Distortions

**Dr. Ravindra Shukl**
Author

**Dr. Kumud R. Bansal**
Translator

**DIAMOND BOOKS**

X-30 Okhla Industrial Area,
Phase - 2, New Delhi - 110020

| | | |
|---|---|---|
| Publisher | : | Diamond Pocket Books (P) Ltd.<br>X-30, Okhla Industrial Area,<br>Phase-II, New Delhi-110020 |
| Phone | : | 011-40712200 |
| E-mail | : | sales@dpb.in |
| Website | : | www.dpb.in |
| Edition | : | 2021 |
| Printed by | : | Adarsh Printers, Delhi- 110032 |
| Design & Type Setting | : | Pooja Rani |

**Varn System : Original Concept and Distortions**
Author - *Dr. Ravindra Shukl (Hindi book)*
Translator - *Dr. Kumud R. Bansal (English book)*

Dedicated to

*Dr. Lal Chand Gupt 'Mangal'*

a brotherly figure for me

## Acknowledgements

When I count blessings and guidance which saw me through this stupendous translation work, my head bows down before Bhagwan Shri Krishn and all astral beings. Some pure souls, residing in heaven, were anchors and did most of the translation work for me. An intellectually frail human being like me could not have completed this colossal task alone in less than 100 days.

I am grateful to Honorable Sh. Kanwar Pal, Minister of Education, Forest Tourism, Parliament Affairs, Art and Cultural Affairs and Hospitality Organization, Haryana, for his accolades.

I appreciate Prof. Deepti Dharmani, a friend of four decades, for her diligent and intelligent help.

Sh. Balbir Punj, Sh. Dinesh Dadhichi, Sh. Jagdish Chopra, Dr. Himmat Singh Sinha, Dr. Ravindra Shukl, Sh. Shailendra Shail and Dr. Shyam Singh 'Shashi' have my gratitude for their pre-publication reviews.

Since my adversaries assist me in my Bodhisattv, a way of life, I long for them like a treasure discovered in the backyard and acquired without efforts. My regards for them.

I am thankful to my friends who nourish my soul.

I am grateful to my incredible family, a source of constant support, love and care.

**Kumud R. Bansal**

# Perspective

"Gentle Lady! What's your name?" An unknown person, walking a few steps behind me asked this question on a warm winter morning of December 1984. My thoughts were interrupted. For me, walking is meditation. It is the time of the day when I want to be just myself. After initial hesitation, I responded to the query. The person introduced himself as Dr. T.K. Parchure. Throughout the morning walk, lasting about an hour, we discussed J. Krishanmurti's book *Awakening of Intelligence*. Both of us were staying at Krishanmurti Foundation, Rajghat Fort, Varanasi. A few days later, Dr. Parchure asked me if I could translate J. Krishanmurti's book *Urgency of Change* into Hindi. I was flabbergasted. Thunderbolt had struck me. Expression of my inability and lack of acumen in the art of translation had no impact on him. Soon he became a father-figure. He taught me the intricacies of translation from one language to another. Under his stewardship, I completed the translation of the book. Its title was *आमूल क्रांति की आवश्यकता (Aamul Kranti ki Aavashyakta).* This book was published by Krishanmurti Foundation in 1986.

Covid-19 virus had made my body its temporary abode. One day, in October 2020, Dr. Ravindra Shukl called me to enquire about my health. Conversation continued for long. Life once again put me to test on the plank of translation as he asked me if I could translate his book the *Varn Vyavstha, Maulik Avdharna aur Vikriti.* We briefly discussed the book. His views were in unison with mine. A seed was sown, which could not sprout till 24.05.2021.

I was conscious that this translation work is a journey down the Hindu scriptural, historical and social fabric of Bharat, which is more than 10000 years old. A joyful play with words, patience, trust and persevarance would be absoultely necessary. I was sure to travel no matter how good or bad the paths were. I was aware that this journey was not going to be a simple journey. And it has been no less than a roller coaster ride.

I humbly request the reader to go on this journey with me. This journey should be with an open mind. Fixed ideas adversely affect a logical, analytical and rational mind. Make a leap out of the dence fog of a cluttered mind and onto upward path. The reader must embrace the winding road and keep the head and heart in sync. Soren Kierkegaard said, "To dare is to lose one's footing momentarily. To not dare is to lose oneself.

Let's start the journey with the journal of my venture.

25 May 2021 - I read the whole book.

26 May 2021 - Wee hours. Captivating shimmer of full moon of Budh Purnima. I was engrossed in scheduling translation of Dr. Ravindra Shukl's book. Suddenly, mind was thoughtless and still. No conscious record of time. I realized the shimmer of moon had given way to glimmer of sun that had penetrated my skin.

28 May 2021 - With each turning page, my head bowed down in awe before the lofty ideal of *Varn System* and *Ashram Dharm* enshrined in Hindu scriptures. The text of the book covers almost 30 centuries. The work of translation could not be free of pitfalls and to make it more slippery the Sanskrit *Shloks* from various scriptural texts spanning many eras are going to be a great challenge.

31 May 2021- Though translation work is proceeding well, yet somehow I feel the fragrance is missing. I introspect. I realize that my work so far has been a mechanical translation done perfectly. What to do?

02 June 2021 - For two days, I keep the book by my side and let it communicate with me. The awful truth begins to dawn on me. I need to be in love with the content of the book not only intellectually but emotionally as well. My heart and mind has a dialogue. The love for the content is now fructifying in translation.

03 June 2021- Difficulties concerning language, structures, idioms and expressions, compound words, two-word verbs, appropriate English equivalents of Hindi words need to be tackled.

04 June 2021- Lexical and semantic problems still stare at me. Several circumstances, instances and persons emerge as powerful catalysts to propel the translation work.

08 June 2021- Questions of temporality, aspectuality and problems of grammar are being befriended.

11 June 2021- Many syntactical problems originate from the point-of-view from the facts are narrated in the book.

12 June 2021- Rhetorical facts, problems like metaphors, comparisons etc. look for solutions.

15 June 2021- Doing the translation work for almost seventeen hours a day was taking its toll on body. Nonetheless mind came to rescue by keeping spirits elevated.

19 June 2021- The most challenging issues I was facing concerned cultural differences and cultural connotations. Neither it could be merely translated from one language to another language nor it could be a simple adaptation based on the author's thoughts.

22 June 2021- I felt the intention of the writer was not to hurt anyone but to unravel the truth, buried deep underneath distorted history books, intentionally rewritten to crumple the grand edifice of Bharat by crippling the *Varn System* based on attributes and *Karm*. Translation aims to bring the truth before English knowing persons.

25 June 2021- I decided that without affecting the intent, objective and aim of the author, a few lines from a few chapters from part II could be omitted. However, this omission would not adversely impact the meaning and sense.

28 June 2021- The book testifies that the way the author has handled the project shows his resilience, experience, knowledge and critical thinking. He has not minced his words while attacking the invaders who destroyed our true Dharmik identity. The quality of his writing and the piercing attack is unprecedented.

01 July 2021- I spent two hours searching *Varn System* on various search engines on internet. Most of the material

found on these search engines is distorted information about everything that has to do with *Bhartiy* Sanatan Sanskriti. Along with Portuguese, Muslim and Mughal invaders and British, these search engines are equal partners in conspiracy. Many theories which damage our ten thousand year old civilization are constantly weaved and uploaded. This literary invasion is fatal. As a result, many generations of Hindu Sanatan Sanskriti have been alienated from their *Bhartiy* roots. The western paradigm has mortally impacted our sacred *Varn System*.

It is difficult not to sink in depression caused by these wrong and distorted informations.

2 July 2021- The warm hues of sunrise have brought a gentle-strong message from nature to boost energy, psychologically and physically. I must chase the goal of translation. The belief in the immortality of *Sanatan Dharm* is renewed.

04 July 2021- Polysemic words and homonymy have to be handled everyday.

07 July 2021- We, as *Bhartiy*, were not enamoured of some other culture or its belief-systems, or living style. It was terror or an instinctive feeling of self-preservation which propelled us to leave our cultural and *Dharmik* way of life.

09 July 2021- One can be in love with the content of the book yet not be in full agreement with it. I think, only those forts fall which are unable to protect themselves from the onslaught.

12 July 2021- The twenty two centuries of horrendous attacks on *Sanatan Dharm*, *Sanatan Sanskriti* and our *Rashtr* are deep injuries imprinted on our consciousness. In my own way, I have tried to heal the lacerating wounds. Present translation work is one of such several ways.

15 July 2021- Wilful adherence to law, wilful deterrence against misconduct, responsible exercise of freedom and liberties and having the societal trait of 'May all be happy' will lead every person, family, society and nation to contentment and perpetual peace. This is to be attained through *Varn System*.

One must appreciate David Frawley for saying, "Hindu *Dharm* is not based upon a religious belief or political ideology, but respect for the living universe and oneself within all, accepts a universal truth behind all worlds and all beings from the physical to the transcendent, not limited to any anthropocentric views."

20 July 2021- It is astonishing that none told us about great kings like Smrat Vikramadity, Lalitadity Muktpida, Samudr Gupt, Kanishk, Kharavela, Rajendr Chola, Mihir Bhoj, Govind III, Pulakesi II, Krishnadev Raya, Mahapadm Nand, Gautmiputr Satakarni.

History books do not mention that horrors of forceful religious conversions and tortures were put on halt by Gurjari Rampyari Chauhan, a 20 years old *Virangana* who led all female regiment of four thousand women and defeated Taimur who, even after winning Delhi had to return to his homeland.

23 July 2021- I regret why rewritten history books Intentionally hide the brave and courageous Rani Durgawati (1524-1564). Why Rani Tara Bai Bhosle, the regent of the Maratha empire of India from 1700 till 1708 and who valiantly fought the Islamic invaders, is missing from the History books?

How many people know about fearless Mata Bhag Kaur, who led a group of 40 Sikh warriors against the 10 thousand strong Mughal army in the battle of Muktsar in 1705? Our school books never teach about Onake Obavva, or Meladi Chennamma, or Belawadi Mullamma.

25 July 2021- It is amazing to know that Smrat Harshavardhan used his scientific skills and knowledge to prepare the first artificial ice pieces. This scientist king was playwright too, who wrote plays in Sanskrit.

How many people from young generation know that ancient Bharat had supreme knowledge about Dentistry (7000 BC), Ayurveda (5000 BC), Ancient flush toilet systems (2500 BC), Ruler (2400 BC), Weighing scale (2400 BC), Plastic surgery (2000 BC), Crucible steel (200 BC)?

For centuries we have been fed with wrong information that in the year 1665 AD, Robert Coach, a botanist with the help of microscope, explained to the world about the plant cell and its structure. 1600 years prior to that date, *Rishi* Parashara in the first century AD, had clearly explained the structure of a plant cell in the Sanskrit work *VRKSA AYURVEDA*

27 July 2021- I take pride in the wisdom of our *Sanatan Rishis* who called this world *Jagat. Gat* means motion. More than ten thousand years ago, *Bhartiys* knew that earth was not stationary; it was rotating. We also knew that earth rotates on its axis and around the sun.

Bhagwan Vishnu's incarnation as *Varah* shows the boar holding earth on the tip of its nose. One may or may not believe in the story about incarnation but the story tells us that earth is round. More than five thousand years ago, we knew that earth was round. Is it not stunning?

30 July 2021- After contemplating about the earth, my thoughts moved into space. I remember my father telling me about the 15th brightest star which our *Rishis* called as **Jayeshth**. This star is called **Antares**. This awesome knowledge of our scientists-*Rishis* still surprises me. Just imagine that 15th brightest star was called **Jayeshth** and not the brighter stars than **Antares**.

In the Southern Bharat, there is a tradition that newly wed couples watch **Arundhati-Vashishth** stars. In twin star system, one star rotates round the other star. In the case of **Arundhati-Vashishth** twin star system, these two stars rotate on their axis and rotate round each other too. One must wonder at the deep knowledge of *Rishis*.

31 July 2021- I feel overwhelmed remembering the skil of the *Bhartiy* artists. Most of the people know about the rust-free pillar in front of Qutubminar. A similar pillar dating more than two thousand years back was built by tribals of Karnataka to welcome Adi Shankrachary.

Four thousand years ago, ancient Bharat knew the technique of distilling Zink from Zink-ore. Countries desirous of procuring Zink had to come to ancient Bharat only. This technique was stolen by Chinese. British stole this technique from Chinese and in the year 1550 a factory of Zinc was established in Britain.

01 August 2021- During a sojourn in the Southern Bharat, I came to know about 'Katapayadi Sankhya'. One of my friends who was a Bhagwan Shri Krishn's devotee recited a *Shlok* in praise of Bhagwan Shri Krishn and asked my views about that *Shlok*. I was a bit puzzled. This *Shlok* was like any other *Shlok* written and sung in praise of Bhagwan Shri Krishn. At this point,

he told me that each alphabet of this *Shlok* carries a particular number of value to it. In the pi theory, the decimal extends into infinity, though it is generally rounded to 3.1416. Rounded to 30 digits past the decimal point, it equals 3.14159265358979 323846264338327 9. In this *Shlok*, if we write the value of each alphabet then we get the value of pi correct to thirty decimal points. I feel amazed, amazed and amazed. We knew encryption too!

07 August 2021- Readers must not think they are reading a history book on ancient Bharat. The reason why I write about spiritual, scientific, technical and military might of ancient and medieval Bharat is that these unbelievable depths and heights could only be achieved by following sacred *Varn System*.

09 August 2021- The word 'secular' means everybody has a right to do what they want to do and others respect it. Secular does not mean you give up everything that belongs to *Sanatan* culture. Secularism does not mean being irreverent to *Sanatan Dharm*.

11 August 2021 - I got up at 3:15 a.m. The very first word which came to my mind was Uttarpara. I could not understand why this word came to my mind. I igonored it. Maharishi Aurobindo's picture kept blinking before me. After sometime, I was able to associate Uttarpara and Maharishi Aurobindo. I searched for Uttarpara and Maharishi Aurobindo on search engines. I read and embibed each and every word of Maharishi Aurobindo's Uttarpara speech of 30.May.1909. I can not resist myself quoting from this speech "We speak often of the Hindu religion, of the *Sanatan Dharm*, but few of us really know what that religion is. Other religions are preponderatingly religions of faith and profession, but the *Sanatan Dharm* is life itself; it is a thing that has not so much to be believed as lived. This is the *Dharm* that for the salvation of humanity was cherished in the seclusion of this peninsula from of old. It is to give this religion that India is rising. She does not rise as other countries do, for self or when she is strong, to trample on the weak. She is rising to shed the eternal light entrusted to her over the world. India has always existed for humanity and not for herself and it is for humanity and not for herself that she must be great."

13 August 2021 - When one ponders, over the downfall of *Varn System* and consequent downward slide of Bharat, one gets caught in the swirl of reasons. Though as a *Rashtr*, we were involved in research and spiritual enlightenment yet our kings valorously, fearlessly and courageously defeated the invaders many times.

We must love *Bhartiy* people's innocence for remaining slaves to even small invaders for many centuries and their infatuation for the Communist Party founded in 1924. Most of the members of this party have betrayed the motherland.

14 August 2021 - We believe in Bharat as one *Rashtr*. The concept of two nation theory is not some modern construct. It is a political variant of the concept of *Ummah* i.e. universal Muslim brotherhood.

15 August 2021- All those who have deep faith and reverence for *Sanatan Dharm* and its scientific basis, must come forward to make synergetic efforts to eliminate distortions, chase away dark clouds of wrong informations, drive off conspiracies of the Left, so called secularists and liberalists and sluggishness of those who believe in *Sanatan Dharm*. I have a firm belief that the clarion call will be heard.

We need another Chanaky, Chandr Gupt and Sardar Vallabh Bhai Patel in our midst.

17 August 2021- I am again reminded of the profound words uttered by Maharishi Aurobindo "It is this religion that I am raising up before the world, it is this that I have perfected and developed through the *Rishis*, saints and *Avatars*, and now it is going forth to do my work among the nations. I am raising up this nation to send forth my word. This is the *Sanatan Dharm*; this is the eternal religion which you did not really know before, but which I have now revealed to you. The agnostic and the sceptic in you have been answered, for I have given you proofs within and without you, physical and subjective, which have satisfied you. When you go forth, speak to your nation always this word, that it is for the *Sanatan Dharm* that they arise, it is for the world and not for themselves that they arise. I am giving them freedom for the service of the world. When therefore it is said that India shall rise, it is the *Sanatan Dharm* that shall be

great. When it is said that India shall expand and extend herself, it is the *Sanatan Dharm* that shall expand and extend itself over the world. It is for the *Dharm* and by the *Dharm* that India exists."

19 August 2021- In the words of Dr. David Frawley "All Hindus should teach their children the fundamentals of *Vedant*, *Karm*, *Punarjanm*, *Moksh*, *Atman*, *Ishvar* and *Brahm*, before they are exposed to negative portrayals of Hinduism in schools. The youthful mind is vulnerable to social influences if not given a deeper understanding."

20 August 2021- A few days ago, I came across a saying that there is no perfection, only beautiful versions of brokenness. I open myself to the impossible and embrace a psychology of possibilities. There are possibilities that reader may find a few grammatical, punctuation or spelling mistakes. My apologies.

21 August 2021- I am only a translator and own no responsibility for the content of the book.

22 August 2021 - Many Hindi and Sanskrit words have no equivalent word in English; even the nearest equivalent word does not convey the cultural and *Dharmik* ethos. Therefore, I decided to retain the original words of Hindi and Sanskrit languages, which will be explained in glossary.

During this journey, reader will be in awe of grandeur, splendour, generosity, majesty, spiritual, scientific and social thinking, contemplations and philosophy of life of Bharat in the first part of the book. In the second part of the book, reader will have to endure the bacchanalias, bibulousnesses, unrestrained orgies; frightening murders, rapes, loots, razing of temples and horrendous crimes committed against Hindus. It will be difficult to digest the anamorphic distortions, cockeyed conspiracies, gnarled attitudes and blatant use of 'divide and rule' policy of the British.

Though the pain and tortures, Bharat has suffered are indescribable, nonetheless, we can not let lose the rope of hope. We must not forget the daggars of betrayals, that our souls were torn and we were objects of scorn. We have to fight with all our might. We have to make a choice for a change to muster our might and boost up our morale. A choice from our hearts and souls to restore the ancient glory of our own Motherland Bharat.I quote from my book 'VIbes' published in 2013.

Arise, awake & stop not,
To be a sheep is not our lot.
Why has the strength in our spine failed?
Where is the spirit with which we first sailed?
What tempests have barred our way?
Why have we lost our vigour, say?
Let us arise, awake, struggle and brave,
Let us defeat the tempest and shackle wave,
Let us strive to reach our destination.
Forward, forward, always forward
Let us march forward, with determination.

We are capable of achieving the goal of making our Bharat a world leader, guide and guru. It is never too late for a new beginning. Let us start following *Varn System*.

Let our efforts not cease,
Let everyone beseech.
A soul full of valour we need,
Now is the time for a lion's deed.
Shake off the shackles of ignorance,
Every temptation and allurance.

This is the perspective emanating from moments of epiphany. Journey should not end with the closing of this embodied text. The end of the book should impell reader to act for rebuilding of *Rashtr* Bharat. I pray the end of the book initiates the commencement of a new era.

Critical comments from readers are welcome.

Samvat 2078, named Anand
22 August 2021
Raksha Purnima

**Kumud Ramanand Bansal**
A devout Sanatan Dharmi

कंवर पाल

शिक्षा, वन, पर्यटन, संसदीय कार्य, कला एवं सांस्कृतिक मामले तथा सत्कार संगठन मंत्री, हरियाणा

दिनांक, चण्डीगढ़ ...............................................

# संदेश

हमारी वर्ण–व्यवस्था, इसकी अवधारणा और उन प्रक्रियाओं, जिनके कारण भारत के इस महान सांस्कृतिक लोकाचार में विकृतियां आईं, पर भारतीय दृष्टिकोण से लिखी गई पुस्तकें दुर्लभ हैं। ऐसी पुस्तकों का अंग्रेजी भाषा में अनुवाद उससे भी दुर्लभ है।

हरियाणा साहित्य अकादमी की पूर्व निदेशक डॉ. कुमुद द्वारा अनूदित यह पुस्तक उन अंग्रेजी भाषी लोगों और युवा पीढ़ियों के भ्रम को दूर करेगी, जिन्हें वर्ण–व्यवस्था और हमारे महान ऋषियों द्वारा स्थापित चार आश्रमों के बारे में सही जानकारी नहीं है।

पुस्तक के लेखक डॉ. रविंद्र शुक्ल ने वर्ण–व्यवस्था के संबंध में अत्यंत बहादुरी से सच्चाई को सामने रखा है। लेखक ने स्पष्ट शब्दों में वर्ण–व्यवस्था की विकृतियों और धारणाओं पर प्रहार किया है।

मेरा दृढ़ मत है कि कुमुद जी ने एक जटिल कार्य निष्पादित किया है। वे मानव मन और अंतर–व्यक्तिगत संबंधों की जटिलताओं की गहरी तथा मनोवैज्ञानिक समझ रखती हैं। आशा है कि साहित्य–सृजन और अनुवाद का उनका कार्य इसी तरह जारी रहेगा।

मैं जीवन में उनकी सफलता की कामना करता हूँ और अत्यंत निपुणता के साथ किए गए कार्य के लिए बधाई देता हूँ।

Kanwar Pal

Minister of Education, Forest, Tourism, Parliamentary Affairs, Arts and Cultural Affairs and Hospitality Dept., Haryana

# Message

It is rare to find books written from Indian point of view on our *Varn System*, its sacred concept and the processes through which distortions occurred in this great cultural ethos of India. It is rarer to see the translation of such books into English language.

Through the translation done by Kumud ji, former Director of Haryana Sahitya Academy, this book will dispel the confusions of English speaking people and of younger generations, who are fed wrong informations about *Varn* system and four *Ashrams* as established by our great *Rishis*.

Dr. Ravindra Shukl, the author of the book has very bravely put forth the truth with regard to *Varn System*. Without mincing his words, the author has directly attacked the distortions and projections in *Varn System*.

I am of the firm opinion that a gigantic task has perfectly been performed by Kumud ji. She is a writer with great depth and psychological understanding of human mind and complexities of inter-personal relationships. I expect that she authors and translates many more books.

I wish her success in life and congratulate her for her stupendous task completed with utmost dexterity.

( English translation of the Hindi massage )

# Pre-publication Reviews

**Balbir Punj**

Translation is an art where the translator preserves the character of original language and yet breathes new life in another language. There is no consensus amongst great authors when it comes to translation; some consider it to be intellectual forgery or think that translation is unfaithful to the original.

While maintaining her own style, Ms. Bansal has retained rhythm and nuances of the author. She has creatively adjusted her tenor with that of the author. Expressed in Sanskrit and Hindi languages, this book carries cultural and religious ethos of more than twenty five centuries,. Often, this ethos cannot be expressed in another language, because that language does not carry the same civilization and cultural background. There are many Sanskrit and Hindi words which will lose their real meaning, music and fragrance if translated into English. By retaining the Sanskrit and Hindi words and explaining them in glossary Kumud ji has kept the spirit alive.

Dr. Ravindra Shukl has brilliantly referred to our *Ved*, *Smritis, Puran* and other religious texts. *Varn System* of ancient India was the social stratification based on persons', attributes, actions and inherent talents. *Varn System*, as originally conceived was not based on birth. *Varn System* was intended to distribute the various responsibilities among various people for the smooth functioning of the society. *Varn System* was totally scientific and based on human psychology, which cannot be ignored even in present times.

Unfortunately, in this wonderful social system, distortions creeped in. These distortions wreaked havoc and edifice of society crumbled down. For whatever reasons, these distortions may have occurred, it is our duty to clear this filth of distortions.

This book is divided in to three parts. The first part of the book has references from original scriptures. The second part of the book deals with Muslim invaders, Mughals and British. The emphasis of the author is on the atrocities committed by invaders and forceful religious conversions. In the third part, the author has identified the problems viz intrusion, flood of Mosques and Madrasas, family planning and Islam. The author has proved the relevance of *Varn System* and have suggested solutions.

There is little doubt that the narrative set by the British was biased, aimed to strengthen their empire and corrode the Indian society from within. No wonder, in spite of intrinsic cultural unity that glues us all, the Muslim league led by Mohammad Ali Jinnah managed to vivisect India in 1947 with the active help from the departing British masters and the ideological underpinning provided by desperate left. Even now there are overt and covert attempts by fringe elements and interested foreign powers to split the society into mutually hostile splinter groups and balkanize the country.

The Indian Left, intellectually nourished by an alien creed, sans roots in the Indian culture, and devoid of connect with local traditions, discovered natural allies in Islamic zealots. The Indian Communist movement believed that India was not a nation, but an artificial aggregate of over dozen mutually exclusive nationalities and programmed to self destruct itself. The Communist leaders and cadres, styling themselves as revolutionaries, took upon themselves the ordained task of accelerating the process of annihilation of Indian way of life and its enterprises, initiated by Colonial masters.

Throughout their cultural war against India, the colonial power received support from a major part of the Church, communists, and later from those who claimed to speak for Islam. Each of the three creeds have one thing in common, all have foreign origins and reject the concept of Bharat Mata, an emotional construct for numerous Indians who continue to pine for a prosperous, united and glorious India of the yore.

Prior to the arrival of the British (first as traders, and later as invaders), India was a huge tapestry, dotted with numerous principalities which lived in peace or at war with each other depending on whims of local potentates. This confusion and uncertainty was not peculiar to India - it was the norm during those times the world over since it was impossible for a central authority to hold on to a large area for long in absence of modern communication and transportation. Germany became a reality in 1871 when 39 small principalities were brought together by Otto Von Bismarck. Before 1861 Italy was a congregation of several kingdoms, duchies, and city-states. And so was France divided in different independent entities or quasi-independent states prior to the 1789 revolution and establishment of French Third Republic in 1870.

Dr Ambedkar, in his last speech on 25 November 1949 in the Constituent Assembly said, "What perturbs me greatly is the fact that not only India has once before lost her independence, but she lost it by the infidelity and treachery of some of her own people. In the invasion of Sind by Mohammed Bin Qasim, the military commanders of King Dahar accepted bribes from the agents of Qasim and refused to fight on the side of their king. Nothing unfortunately has changed so far. We had traitors then, we have them now as well. Dr. Ambedkar was prophetic, his words still hold true!

Most of the so-called *'Dalit'* and Left outfits propagate a manufactured narrative which runs something like this: 'Caste Hindus are Aryan invaders who subjugated the original inhabitants (Dalits); *Brahmans* wrote Hindu scriptures justifying inhuman treatment to the "aboriginals" and are responsible for their plight'. This construction is flawed and mischievous as well. The demonization of *Brahmans* is without any basis, they did not compile or compose any major religious work. The *Veds* and the *Mahabharat* were put together by *Ved Vyasa*, son of a fisherwoman. The *Ramayan* was penned by Valmiki, revered as Adi Kavi and the community named after him, is now classified as *Dalit*. Manu, the author of the *Manusmriti*, was not a *Brahman*.

Since the Hindu society, according this distorted paradigm, is inherently unjust, *Dalits*, should desert or destroy it to improve their lot. The options available to *Dalits* in this sad scenario are: either convert to one of the Abrahamic faiths or turn non-believers-Communists. Dr B R Ambedkar, (most of the *Dalit* organizations swear by him) dismissed this motivated construction completely.

In his paper "Caste in India", Dr. Ambedkar wrote: "one thing I want to impress upon you is that Manu did not give the law of caste... caste existed long before Manu". He found no reason to blame *Brahmans* for caste system. "The spread and growth of the caste system is too gigantic a task to be achieved by the power or cunning of an individual or of a class. Similar in argument is the theory that the *Brahmans* created caste. After what I said regarding Manu, I need hardly say anything more, except to point out that it is incorrect in thought and malicious in intent... *Brahmans* may have been guilty of many things, and I dare say they were, but imposing of the caste system on the non-*Brahman* population was beyond their mettle".

Dr. Ambedkar didn't buy the Aryan invasion theory either. On the origin of *Shudrs*, he said: "The *Shudrs* were one of the Aryan communities of the Solar Race. There was a time when the Aryan society recognized only three *Varns*, namely *Brahmans, Kshtriys and Vaishys*. The *Shudrs* did not form a separate *Varn*. They ranked as part of the *Kshatriya Varn*. There was a continuous feud between the *Shudra* kings and the *Brahmans* in which the *Brahmans* were subjected to many tyrannies and indignities".

Dr. Ambedkar further says: "As a result of the hatred towards the *Shudrs* generated by their tyrannies and oppressions, the *Brahmans* refused to perform the *Upanayana* of the *Shudrs*. Owing to the denial of *Upanayana*, the *Shudrs* who were *Kshtriys* became socially degraded, fell below the rank of the *Vaishys* and thus came to form the fourth *Varn*".

The Constitution-makers in a bid to protect minorities from majoritarian interference had inserted a series of Articles. What was meant to prevent discrimination against minorities has in fact turned out to be an enabler of discriminatory treatment by the state against the majority.

The present work by Dr. Ravindra Shukl is an effort to understand the vexed issue of *Varn System*. The author has put the subject in a proper perspective and tried to unravel the mischief which is being carried out in the name of *Varn System*. Of course, translation has its limitations. Still, Ms. Kumud R. Bansal has done a brilliant job. Congratulations to both the author and the translator as well.

Contact : punjbalbir@gmail.com Former Rajya Sabha member

**Prof. Deepti Dharmani**

Having watched Kumud ji translating a significant, scholarly socio-historical treatise is a veritable lesson to me into the gestation and labour a writer/ translator undergoes for delivering an embodied text. She aims at nothing less than perfection and meticulously executes her plan with perseverance and utmost dedication. Writing a book demands a mammoth effort and translating it requires no lesser. This is especially when the book takes up an issue long held under controversy with a malignant design doggedly followed it. To allude to WB Yeats the best have lacked conviction and the worst have been full of passionate intensity. For the translator the obstacles multiply owing to cultural and linguistic reasons. I can assuredly say that Kumud ji has the caliber to surmount both.

Since therein lies a chaste purpose behind translating the text, I wish the book gets a wider readership and the readers are able to peep into the past, fathom some nuggets of wisdom and relook at and grasp the original concept of *Varn Vyavastha* that was rooted in the inclusivity of humanity, and productivity and perfectionism. It was as much spiritual as it was social and economic.

The book is argumentative in nature, and hence demands no unquestioning acceptance of the truth. With all the citations as documentary evidences, the book explodes the misconceptions as well as the nefarious schemes of the enemies of Bharat existing both within and without. Those who were driven by passion distorted its original spirit. Still such people as are propelled by Raj (passion) may continue the debate endlessly but those who value and seek Truth, those whose nature is conditioned with goodness, generosity and human candour and pursue the egalitarian philosophy will be able to accept the 'difference' beyond sectarianism and orthodoxy.

Saint Tulsi Das epigrammatically summed up the entire philosophy of action in just two lines in the *Ramcharitmanas*: *जाकि रही भावना जैसी, प्रभु मूरत देखी तिन तैसी।* This statement is all about human perspective towards life. It is attitude which defines a person's behaviour.

Contact : Dharmani749@gmail.com Former Dean of Academics
Chaudhary Devi Lal University, Sirsa

**Dr. Dinesh Dadhichi**

Your "Perspective" is so well crafted that I went through the contents in a single reading. Your expression is flawless.

In this introductory note, you have employed two methods. At the outset, there is a narrative that arrests the reader's attention. This is followed by the method of revealing the different stages of your work through a date-wise journal.

Both these methods involve the reader and invite him to participate in the intellectual experience so beautifully described by you.

Translation of a text that is fraught with quotes from ancient Sanskrit texts into English is a formidable challenge in itself. Evidently, there are cultural differences and in several cases the translator has to innovate a great deal. It involves a lot of creativity. You are well versed in the art of translation, and I am sure your work will prove your acumen.

Appropriate, interesting and powerful as a foreword!

Contact : dk.english@gmail.com Former Professor and Chairman,
Department of English,
Kurukshetra University, Kurukshetra (Haryana)

**Dr. Himmat Singh Sinha**

The present book is a laudable English translation of a Hindi book by Ms. Kumud Bansal, a meritorious Hindi scholar, on a much maligned issue of social stratification called *Varn Vyavastha*. It is a hazardous task to conserve the exact sense which the original words, phrases, idioms in classical texts convey because there are hundreds of Sanskrit words which have no exact equivalents in English. A translation to serve its purpose must be as clear as its substance will permit. But it is the marvelous quality of the present translation that the learned scholar Ms. Kumud Bansal has no where allowed the purpose of the original to get distorted. This proves her command over both the languages and her mastery over translation skills, which shall be appreciated by the able readers.

So far as the issue involved is concerned, the *Varn System* was evolved very wisely by Indian seers and sages for saving the society from all external and internal dangers, social degeneration and petrification. In the *Gita,* it is said that the four-fold division of society into intellectual class (*Brahman*), the warrior class (*Kshtriy*), the producer class (*Vaishy*) and the unskilled labour (*Shudrs*) was created by the Bhagwan himself according to quality and work. Here, in the *Varn System*, the emphasis is on '*Gun*' (aptitude) and *Karm* (function) and not on birth. A class determined by temperament and vocation is not caste (Jati). This four-fold division is designed for human evolution. Such functional grouping is essential for the all round progress of society and preservation of civilization. In her translation the wise translator Ms. Bansal has very beautifully retained the grace and purpose of the original by selecting appropriate words for the Sanskrit words and verses. This difficult task has been accomplished by Ms. Bansal with great excellence, which reflects that the scholar has deep understanding of the subject-matter.

Language of translation is simple, free form unwarranted pedantry and can easily be grasped even by a lay reader. The style is very lucid, which is hallmark of fluid exposition.

I feel confident that the learned readers will receive the book with welcoming spirit. The book shall not only provide intellectual satisfaction but shall also provide the English-knowing people the way to have deep knowledge of Indian social system and Indian culture.

Contact : 80530-49390

Ph.D.

Former Chairman, Department of Philosophy

Kurukshetra University and

Former Member of Indian Council of Philosophical Research

**Jagdish Chopra**

A scientific and perfect *Varn System* was well-devised by our sages. Lamentably, this was not only distorted but disastrously misunderstood too.

Despite the life order being arranged for all *Varns*, many distortions took place, which led to the belief that *Varn System* is based on birth. This is not true, *Varn* is not lineage based. Various *Varns* are but categories determined by the individual's conduct. Dr. Ravindra Shukl, in his book in Hindi has boldly brought forth the truth regarding our scriptures. Dr. Kumud has done the intricate work of translation of this book in English. Literary persons should take translation work seriously. This is the only method one can know the literature of another language. The book *'Varna* system *: Original Concept and Distortions'* will be well-received by the English speaking people.

Best wishes to the translator and the author.

Contact : jchopraadv@gmail.com

Former Political Advisor to

Chief Minister of Haryana

**Dr. Ravindra Shukl**

To dispel the thick veil of darkness of hatred, it is neccessary to make people aware of our scriptures. To acheive this purpose, such like books should be translated into various languages so that the speakers of different languages can become familiar with the basic concept of the *Varn System* as prescribed by the Indian scriptures.

Ms. Kumud Bansal's translation of this book into English is spontaneous and lucid. It will be a reference book for the present generation and people speaking English language. I have no hesitation in saying that she did not receive any help from me in the work of this translation. Nevertheless, she has done a great favor to the present and future generations of India by providing quality translation for the welfare of India. I applaud her good efforts and express my gratitude.Pray to the Lord that she lives long and continues to do welfare work like this. Astu! *Shubham Bhuyat!*

Contact :
kaviravindrashukla@gmail.com
hindisahityaBharati@gmail.com

President,
Hindi Sahitya Bharati (International)
Jhansi

**Shailendra Shail**

I am impressed by the English translation of Dr Ravindra Shukl's thesis in Hindi, rendered by Dr. Kumud Bansal under the title *"Varn System: Original Concept and Distortions"*. Translation is an art and a difficult one. There is a difference between translation and transliteration. Paul Goodman, an American author has said, "To translate one must have style of his own, for otherwise the translation will have no rhythm or nuance, which comes from the process of artistically thinking through and moulding the sentences; they cannot be reconstituted by piecemeal imitation." I think Dr. Kumud Bansal has passed this litmus test while translating this book.

For Dr. Kumud Bansal, this is a labour of love. She is a keen student of Philosophy and has equal command over Hindi and English. Her deep understanding of the subject has resulted in conveying the essence of the original text in Hindi. There are words with subtle cultural nuances in every language. She has been able to retain the flavour and fragrance of the original text in her translation. It is good that she has retained the original

Hindi words wherever she could not find suitable words in English due to cultural differences.

As a translator, one has to be as true to the original as possible. Dr Kumud Bansal has successfully accomplished this. I convey my heartiest congratulations to her.

Contact : shailmohan@gmail.com Editor, AFA NEWS (EAGLES EYE)
Member, Governing Council,
The Poetry Society (India)
Patron, UDBHAV, Cultural and Literary Association

**Dr. Shyam Singh 'Shashi'**

Dr. (Ms.) Kumud R. Bansal is basically a poetess of English. In 2013, I wrote a foreword for her impressive anthology 'Vibes'. I was delighted to read Kumud's poetry in the past. I am equally delighted to go through the script of her translation of a book, dealing with a variety of topics, viz great and sublime philosophy, culture, *Varn Dharm*, Vedic-literatures, the *Quran*, interpretation of the *Manusmriti*, reframing of old history, Muslim invaders and rulers etc. Dr. Ravindra Shukl has studied man and society from an integrated approach of old Hindu scriptures and *Purans* etc., quoting various thoughts and experiences. The extreme tortures inflicted by Portuguese, Muslim invaders and Britishers are not only recorded in the history books but these are duly preserved in various archives. These dark pages should be studied by Indian scholars in depth as Dr. Ravindra Shukl has done in his book. I hope, the author of this book and the translator, will write another monumental book carrying 3000 years of history.

When Ms. Kumud approached me to write a few lines for her translated work, I asked her a question, "What English word have you used for *Prakshepit*?" She told that she has used the word 'interpolations'. I cherished her use of the word 'interpolation' instead of commonly used word 'projection'. I appreciate and acknowledge the intelligent hard work of Ms.

Kumud in translating this book. I admire her deep understanding of the subject of the book and her command over both Hindi and English languages.

It may seem a bit of diversion, but it necessarily supports this book's content. Caste system should be studied in the context of *Gotr* system, as similar *Gotras* are found in various Hindu communities. This will affirm the principle of migration from one *Varn* to another *Varn*. This will eliminate the social caste system and make the country fulfill the dream of *Vishnupuran*, which, too is the dream of Dr. Ravindra Shukl and Dr. Kumud Ramanand Bansal.

गायन्ति देवाः किल गीतकानि धान्यास्तु ये भारतभूमिभागे।
स्वर्गापवर्गास्पदहेतुभूते भवन्ति भूयः पुरुषाः सुरत्वात्।।

विष्णुपुराण 2/3/24

*Gods sing songs that people born in Bharat are luckier than Gods. Bharat is the path to Heaven, Moksh, Kaivlaya. To born in Bharat is better than attaining Godhood. (Vishnupuran 2/3/24)*

Swami Vivekananda has rightly said in his book, "The debt which the world owes to our Motherland is immense. Taking country with country, there is not one race on this earth to which the world owes so much as to the patient Hindu, the mild Hindu. To many, Indian thought, Indian manners, Indian customs, Indian philosophy Indian literature, are repulsive at the first sight; but let them persevere, let them read, let them become familiar with the great principles underlying these ideas, and it is ninety-nine to one that the charm will come over them, and fascination will be result"

Dr. Ravindra Shukl, the original author and Dr. Kumud R. Bansal, the translator, have unindebted themselves from the debt of the motherland Bharat. Congratulations to both.

Contact: B4/245, Safdarjang Enclave, Ph.D. D. Litt.
New Delhi - 110029 President of India awarded Padmashree
shyamskha@yahoo.com for Hindi and English literature

# VARN SYSTEM

## Original Concept and Distortions

# Index

## Part-I

Assertions... ....37
Foreword... ....41
Glorious Past... ....45
Sublime, Philosophy and Culture ....47
Spiritual Vision of Bharat ....53
Dharm : Etymology and Scope ....54
Varn Dharm ....57
Ashram Dharm ....59
References from Scriptures ....64
References from Purans ....77
Bhagwat Puran ....80
Bhavishy Puran ....84
References from Upnishads ....90
References from Smritis ....94
References from Mahabharat ....102
References from Gita ....110
References from Ramayan ....115
References from Jainism ....139
References from Buddhism ....143
References from Sikhism ....149

## Part-II

Distorted form of system ....161
Distortions in the Original Varn System ....164
Tyranny of Mughals ....197
British Conspiracy ....201

## Part-III

Problems and Solutions ....207
Glossary ....225
Bio-Datas ....230

# PART-I

# ASSERTIONS

Driven by the desire to rule over Bharat permanently and extensively the Britishers adopted the policy of divide and rule. On the one hand, they brought about radical changes in Her education policy to root out their sense of pride as *Bhartiys* and on the other hand the Britishers distorted Her history so that besides corrupting its culture, the sense of oneness and symbiosis is ripped apart from the psyche of the coming generations. While they strategically changed the history and education policy of Bharat, they divided the *Bhartiy* society on caste lines. They instilled in the hearts of the so-called low caste people that the upper caste people have grossly oppressed and exploited them and that the treatises written by *Brahmans* are also responsible for their misery.

Britishers, by hook or crook, wanted to destroy the sense of pride of *Bhartiys* and debilitate Bharat by creating caste animosity. They were able to achieve it to some extent.

Strategically Britishers transferred the power in the hands of such persons as would help them materialize their long-term plan. Consequently, caste disharmony instead of depleting kept increasing. Definitions of *Bhartiy* literature had been distorted long before. Distorted explanations were given by the so-called western scholar Max Muller as well as by some such progeny of Bharat Mata as either had a diseased mind-set or were driven by vested interests to weaken Bharat. Instead of putting an end to casteism in the 74 years of Independence, our political parties and governments have kept encouraging it to feed their vested interests, and common people keep following them by becoming a part of their conspiracy. Even the venerable Mahatma Gandhi, who had the power to decide the direction and condition of Bharat at the time of Independence, could hardly see through it, though he was not a part of this conspiracy. Although, he made attempts to fill this trench of caste disharmony, yet he could not voice his protest against this conspiracy to the extent and with force that was required. With piety he addressed the backward section of society as *Harijan*. Over the period, political powers used the term *Harijan* as a weapon. Gradually, the word *Harijan* got replaced with the word *Dalit*. A well-planned action was carried out in the name of *Dalit* thought to cement vitriolic design in the psyche of this section. To make the squid feelings permanent in hearts of this class, well planned things were being done in the name of '*Dalit* Deliberations'. No stone was left unturned to create caste animosity. However, due to the strength of the *Bhartiy* philosophical thought, horrors of caste conflicts have been averted, else things would have been in ruins. Rich cultural heritage of Rome, Sparta and Greece have vanished completely, but the conspirators have failed to fully destroy the *Bhartiy* culture and amity.

It is pertinent to mention here the fact that there is an organization working amongst *Dalit* officials and employees in Bharat. In a well-planned way this organization has worked for creating rifts among the government officials and employees and thereby has entrapped the minds of common people in the cobweb of caste distinctions so that these people like puppets dance to their tunes and could be used as their private property to achieve their political ambitions.

I have a friend who is infused with a sense of fanfaronade and is one of the office bearers of this organization and who once animated with self-praise unveiled a secret to me that the founder member of this organization, once while addressing a gathering of the officials, had retorted that they could not have achieved the status they have achieved today, had they not abused *Brahmans*, *Kshtriys* and *Vaishys* in full throttle voice and held *Bhartiy* scriptures responsible for poor condition of *Dalits*. Could the *Dalit* Community be so consolidated and aggressive without it? We can very easily understand the latent intentions from the unveiling of this secret.

Through a bird's eye view of the evolutionary history of caste rancour we will easily understand that there has been a well-planned conspiracy to alienate us from our own backward brethrens, though during the Islamic rule in spite of suffering and torture most of the people did not change their *Dharm*. At that point of time, they all belonged to higher strata of society, but during Islamic regime, through administrative orders they were made untouchables. Currently a conspiracy is in progress to separate these people from the mainstream society. *Manuwad* has become an abuse. In the castist storm *Rashtr* has been marginalized. People who claimed themselves to be *Rashtrvadi* also started talking of caste equations. Casteism was openly advocated through rationalist discourses. In this race of competitive casteism even the great personalities of the *Rashtr* were shackled and put into caste penitentiary. Caste-based organizations mushroomed every day. Even sub-caste shops came to sell themselves in this caste market. The so-called higher strata did not lag behind.

Nonetheless, it is also a hard fact that groping in the darkness of ignorance and avarice, the scholars from this so-called higher strata left no stone unturned to nurture and embosom the poisonous creeper of casteism. During the British rule, a section of such anti-social, ignorant, and passion-driven people committed inhumane atrocities on the so called lower strata of people but this is not a generalized truth. This was, thus, publicized. In fact, the truth is that only a handful of the anti-social and selfish, and the so-called scholars with evil intentions deliberately gave morbid explanations of Hindu scriptures on the basis of which venomous separatism was fanned in society. Yes, it is also true that there are some social organizations that engage themselves to antidote the poison but the effect of this poison instead of diminishing is still spreading more and more, which is a matter of concern. Probably due to the deep intensity of the negative thinking, positive thinking is not showing effect. The thesis *Varn System : Original Concept and Distortions* is the product of this contemplation.

I have tried to incorporate in this book the original concept of the *Dharm* according to *Bhartiy* scriptures as well as the distortions thereof. But it is not the ultimate. According to my competence, I have made an effort to put the essence of scriptures before you. I am indebted to Mahakavi Awadhesh ji and Shri S.S. Tiwari (retired lecturer of Sanskrit), who during my research not only motivated me every day but they also provided me their able guidance. I am thankful to my friend Pawanputr Badal ji, Manager of Lokhit Prakashan, who understanding the significance of this *Granth*, got it published and acted as a bridge to make it reach your hands.

Astu! I am submitting this *Shodh Granth* in your pious hands at a time when casteism is fully determined to destroy the micro fabric of the society. This *Shodh Granth* will certainly prove to be a *Mantr* for re-establishment of Truth and worshipers of humanity to detoxify the society of casteism.

The need of the hour is to put before the average person the original scriptural concept so that they understand the piousness and serenity of *Dharm* and also know that foreign powers as well as vote-bank politics are responsible for their misery and not our *Rishis*, *Munies* and religious scriptures.

Although, I do not consider myself a scholar, yet I presented whatever I could collect as per my efforts and competence. I am not at all hesitant to proclaim that conspirators have polluted our scriptures, may be in a disguised way or through wrong explanations. Despite it, there is still a lot in our scriptures that can enable us to understand the original concepts of *Rishis* and *Munies*. It is high time that truth-seekers throw away the interpolated parts of scriptures and bring forth the correct explanations to stimulate society. So that we leave the legacy a clean and prosperous society along with a strong philosophy for the forthcoming generations. To achieve this end this *Granth Varn System : Original Concept and Distortions* is a small effort in this direction.

Astu! your blessings will surely make this *Granth* an accomplishment. I am sure that it will demolish the caste-prison and the sun of a new consciousness will arise. This is my humble entreaty to the *Paramatma*, the Supreme Consciousness.

May there be auspiciousness.

Yours
Ravindra Shukl 'Ravi'
Former minister of state
(With Independent Charge)

# Foreword

Once upon a time, Bharat was the crown of this world but at present, it is in a pitiable situation. Bharat is gradually moving towards slavery, unaware of what misfortune is waiting for Her on the path on which she is heading. Knowingly or unknowingly, Bharat seems to be eager to let her pride turn into ashes in the blazing flames of conspiracy. Illiteracy and ignorance are adding fuel to fire. Power-hungry political parties and a few political leaders sold out to the tinker of Euro & Petro, are playing a dirty game. In these circumstance, it would not be surprising if Bharat almost vanishes in the blue.

On the one hand this sponsored Sursa of casteism is ready to swallow the whole society and on the other hand dirty dance of prosylitization is devouring the very roots of Hinduism. Resultantly the bright lighthouse of the world is losing its identity. Unfortunately vote-tempted and power hungry political leaders like Jai Chand, Man Singh, Mir Jafar and Dulhayu are using illiterate and ignorant public and are acting as agents in this nefarious plot. It is shocking that most people are silently watching these terrifying dangers and knowingly or unknowingly are becoming a part of this conspiracy. The feeling of 'कोऊन नृप होय, हमहिं का हानि' *(who so ever may be the king, no harm to us)* has paralyzed the majority society and down fall has gained speed.

This very mind-set tightly gripped Bharat into a slavery. A handful of invaders on ponies, donkeys, and horses were able to reach the Somnath temple. They were able to trample down and plunder the riches of Bharat and return safely. A little bit of resistance was there. Had there been unified resistance, neither *Bhartiy* could have been enslaved, nor the invaders could have returned safely from Bharat. Due to long subjection, impressions of slave acculturation got deeply embedded in our hearts. Alien impact on the mind is also visible on our contemplations. Obviously, during the slavery period we lacked the courage to assert that we were best in world; our *Dharm* and our civilization was unique in the entire world. Now, we are independent. What compulsions are stopping us to declare that we are best? What reasons are stopping us from making this open, loud and frank declaration before the world? Why are we not affirming the fact that the Hindu *Dharm*, Hindu thought, Hindu civilization and Hindu culture is unmatched in the world?

Hindu *Dharm* is the only *Dharm* in the world where peace can be realized. We are not only unable to state so but are also unable to affirm that Hindu *Dharm* is the oldest. We are unable to garner courage to say that when people all over the world were leading a tribal life and did not know the art of living, Bharat, had already timeless *Veds* with the firm declaration of 'कृण्वन्तो विश्वमार्यम्' *(let the world be super excellent).* To light the flame of knowledge we were travelling all over the world. The very name Bharat is divulgement of this truth. 'Bha' means knowledge or light, 'Rat' means seeped in, implying that which is seeped in knowledge or is in search of Truth. Due to frustrated and defeated mentality today's scholars, writers and intelligentsia do not want to read the glorious pages of our history, and by ignoring the past and fabricating baseless and irrational stories are spreading venom in society. Consequently, Bharat, that was excellent in commerce, craft, knowledge, and technique is not only becoming weak but She also seems to be helpless and unequipped.

It may be the play of destiny or conspiracy that sons and daughters born from the womb of Bharat Mata are not faithful to Her. Instead of eliminating the destructive notions, emboldened and instigated by aliens, our own people are spreading falsehood. Due to petty, vested motives they are openly cursing the super excellent *Dharm*, culture and philosophy. Being unaware of truth, the *Dalit*, exploited, victimized and deprived class holds the Hindu *Dharm* responsible for their condition. On the one hand, the Christian prosylitization is blooming whereas Islamic religious frenzy is fast devouring the *Bhartiy* pride and identity. Bharat, whose bright and honoured flag was high in sky from Kaba to Koria and from North pole to Indian Ocean, seems to be subjected to abjectness and misfortune. It is jinxed and regrettable. This situation should be pondered over by every *Bhartiy*.

When the Ganga overflows its banks, the villagers or people living in towns on the shores of the Ganga do not realize that flood has softened ground underneath. Life moves on normally. Suddenly village after village, get encaved into this hollowness. Likewise, many human dwellings start appearing on the cooled-down volcano. One day eruption of volcano destroys everything in a moment. Similarly, misfortune has silently engulfed Bharat from all sides. Islamic terror and dream of converting the whole of Asia to Christianity in 21st century is blindly but in a determined way have accelerated prosylitization. People with vested interests, directly or indirectly, are protecting this morbid mentality. This abhorrent casteism is a grave problem for Bharat. If not diffused within time, we are bound to lose everything. Like Roman, Spartan, and Greek cultures we will be found in the pages of history books. Our existence will come to an end.

*Rashtr* Kavi Dinker said :

जिसको न निज गौरव तथा निज देश का अभिमान है,
वह नर नहीं नर पशु निरा और मृतक समान है।

*That person who has no self-respect and is not proud of his country is not a human being but is like an animal or equal to dead.*

It is unfortunate that instead of realizing this, the present generation is totally ignoring it. Trash of western thoughts is like butter to them. Previous generations are more to be blamed. Well-planned conspiracies were hatched to demolish the sense of pride of Bharat. In spite of creating the awareness about the planned conspiracies, generations after generations did nothing to find solutions. Muslim invaders, in the name of Islam, not only trampled down Bharat but they also committed genocides to destroy *Bhartiy* culture. Jajia Tax, which was almost sixty percent of the crop was imposed only on Hindus. Those who could not pay the tax either had to accept Islam or had to become slaves. The properties and kingdoms of the defeated kings were confiscated by the Mughal rulers. The sepoys, wives, girls, and children of the defeated kings were enslaved. As per the directions of the *Quran*, they were considered *Mal-e-Ganimat* (looted property) and were consumed accordingly. Those races which put straight fight were either enslaved or were ousted from public life and were not allowed to take water from the public wells. This is how untouchability started. A few thinkers used this fact to divide Hindu society into castes. With great effort, this urn of untouchability, was malafidely broken on the heads of *Brahmans*. *Bhartiy* scriptures were also held responsible for it. This malicious campaign was continued by the Britishers. *Manas-Putr* of Macaulay did not lag behind. Vote bank politics in Independent Bharat, through spiteful propaganda, put open oblations into this disgusting conspiracy that divided Hindu society.

The Hindu *Dharm*, Hindu culture, Hindu civilization and Hindu philosophy is supreme in the world. But Hindu progeny fails to recognize its importance. This is the reason why the present generation thinks that *Bhartiy* philosophy, culture and civilization is rotten, decayed, and outdated. Woefully, this notion has got the support of the *Manas-Putr* of Macaulay, who call themselves intellectuals.

Swayed by the materialistic world no one has time to know, understand and remember the glorious past. It is lamentable they echo whatever they hear. At the call of one jackal many jackals join. Together they create horrendous noise amidst which 'Truth' becomes a causality. Due to the numerical strength of the antagonists most of the intelligentsia who know the truth are unable to roar back. Those who want to reveal the truth, even in hushed voice, are compelled to bow down to the so called intelligentsia. Need of the hour is that glorious past should reach to every person so that the sentiment of *Rashtriy* pride sprouts in the psyche of common man.

Hindu Samrajya Dinotsav
Samvat 2079, Yugabd 5116
11 June, 2014

Shridhar Pradkar
*Rashtriy* National Sangathan Mantri

# Glorious Past

The ancient and pristine culture and philosophy of *Bharatvasrh* cannot generally be confined within time framework. There was a competition to underestimate *Bhartiy* culture. This planned conspiracy started from the Mughal period and continued till the British era. If we put all the evidence collected from research thesis available within and outside the country, the picture of a great *Bharatvarsh* comes before mind. According to the depiction of Bharat in epics etc., Afghanistan, Iran, Iraq, Indonesia, China, Japan were within the boundaries of Bharat. Gandhari, the mother of Kauravas and Shalya, who were from Afghanistan, were important characters in the *Mahabharat*. Iran was originally Arya Land. Its last king Rja Shah Pahlavi was more of a follower of the Aryan principles than of Islam. The *Zend-Avesta*, the holy book of Parsis is much like *Rigved*. Burma (Myanmar) was our *Brahm* land. There is a reference of *Iravat* in the *Mahabharat*. The present day *Iravat* velley's connection with the *Mahabharat* is, thus, evident. Sri Lanka in the south was never considered to be separate from Bharat. The Himalayas in the north, Aryan i.e. Iran in the west, Shringpur i.e. Singapore in the east and Sri Lanka in the south can be seen in the map of Bharat year after year.

Ramlila of Indonesia is world famous. The rulers and residents of Indonesia have neither Islamicised their names nor have disowned their ancestors, so much so that buildings, avenues, or areas are named as per the Aryan traditions. You will be surprised to know that Indonesia has no Ambassador in Pakistan. People may not know the reason behind it. Initially when Indonesia appointed an Ambassador to Pakistan, Islamic fundamentalists objected to his appointment due to his closeness to the Hindu culture and civilization. When Indonesian Ambassador celebrated Vijayadashami festival, the government of Pakistan reacted sharply. The annoyed

Indonesia recalled its ambassador saying, they had changed their sect, not their ancestors. After perusing the recent research, it becomes known that our ancestors ruled over Japan. From the available evidences it is proved that Balram, brother of Bhagwan Shri Krishn had established his kingdom in Japan.

Pakistan, Bangladesh, Nepal, Shri Lanka, Burma, Singapore etc. were parts of Bharat before independence. From this, it is evident that geographical boundaries of the great Bharat were extremely vast as compared to the present Bharat. It is a proof of the glorious past of Great Bharat that had a dignified culture.

## Sublime Philosophy and Culture

Many references are found in *Bhartiy* scriptures regarding the eminence of Bharat, such as :

दुर्लभम् भारते जन्म मानुषम् तत्र सु दुर्लभम्।

*It is rare to be born in Bharat and rarer to be born as a human being.*

गायन्तिः देवा किल गीतकानि, धन्यास्तु ते भारत भूमि भागे।
स्वर्गापवर्गास्पद हेतुभूते, भवन्ति भूयः पुरूषः सुरत्वात्।।

Intendment of this *Shlok* from the *Vishnu Puran* is that in the whole world *Bharatvarsh* is the only place where even Gods are eager to take birth. What makes Bharat unique in the world is its marvelous and pristine natural environment, roaring rivers, adorning six seasons, polarity of dialects, costumes, food habits and the wonderful sense of unity in diversity. Swami Vivekanand said, "If there is any auspicious land on the earth which can be called virtuous, where every *Aatma* that is moving on the path leading towards God has to come as a last resort, is Bharat. Of course this country is the chosen one for divine-realization."

Long time ago, a German monk came to Haridwar and chose his death by offering his body to the holy Ganga. One of his letters got published in some newspaper. He had written, "I am myself renouncing my body, offering my body to pious waters of the Ganga, so that with its blessings I will be fortunate to be reborn in Bharat. With new immaculate body, I will be capable of divine-realization."

After a deep study of all the cults, creeds and sects we can easily conclude that the founders of none of these had perceived God in person. With reference to great saint Jesus Christ it is assumed that only angels met him. He met Satan only once. When he was being crucified, for a moment, he had doubts about the compassion of God. Due to extreme pain and anguish he uttered "Oh my God! Why did you leave me?"

Hajrat Mohammad, the founder of Islam only met Archangel Gabriel and heard some divine voices.That is why all

*Bhartiy Vedic Rishis* addressed human beings as *Amrit-Putr* and in solemn and demure voice proclaimed :

वेदाहमेतं पुरूषं महान्तम्, आदित्यवर्ण तमसा परस्तात।
तमेव विदित्वाति मृत्युमेति नान्यः पन्था विद्यते ऽयनाय।।

*I have realized this great being who shines with effulgence like the sun beyond all darkness. One is beyond death only on realizing him. There is no other way to escape from the circle of births and deaths.*

There is no other concept in the world-literature equal to the above concept of self-credence and self-cognition.

Even after immersing himself into all philosophies, the modern and inquisitive student Narendra was not satisfied. At someone's suggestion he met Saint Ram Krishn Paramhans and directly asked : "Sir, have seen God?".........

.............Without a moment's hesitation that Saint of Bharat spoke thus "....... Yes! I see him as I see you. I can show him to you also." And he did that. This very Narendra after divine-realization came to be worshipped as Swami Vivekanand.

In the *Shrimad Bhagwad Gita*, Shri Krishn himself assured every *Bhartiy* :

यदा यदा हि धर्मस्य, ग्लानिर्भवति भारत।
अभ्युत्थानमधर्मस्य तदात्मानम् सृजाम्यहम्।। 4, 7

*Whenever there is a decline in righteousness and an increase in unrighteousness, at that time I manifest myself on earth. (4,7)*

Acharya Tulsi supports this declaration in the *Ramcharit-manas*:

जब जब होय धर्म की हानी। बाढ़हिं असुर अधम अभिमानी।।
तब तब प्रभु धरि विविध शरीरा। हरहिं कृपा निधि सज्जन पीरा।।

*Whenever Dharm is in danger, when demons and conceited people grow in number, then God takes many forms and with his blessings, pain and sufferings of noble people vanish.*

It is clear from these references that God incarnates only on the land of Bharat. There is a long chain of Divine incarnations in Bharat. Equally long is the chain of *Rishis*. It is difficult to remember all the names.

For Christians the holy book Bible and for Muslims the Pak-e-*Quran* is the ultimate ideal, whereas *Bhartiy* Saints clearly declare: 'मुण्डे मुण्डे मतिर्भिन्नः' *(every person has his own opinion)*, 'एकम् सत् विप्राः बहुधा वदन्ति ' *(truth is one. the Rishis tell this truth in various ways).* In fact, there is no equivalent word formation in English to express beautiful and classical pronouncement such as 'यत् पिण्डे तत् ब्रह्मण्डे' *(whatever is in body, is in universe)* 'अहम् ब्रह्मास्मि' *(I am the universe)* etc. These are also a clear reflection of the *Bhartiy Rishis*' direct realization of God. The greatness and excellence of Bharat is also indicated in the following verse:

एतद्देश प्रसूतस्य शकाशादग्रजन्मनः ।
स्वं–स्वं चरित्र शिक्षेरन पृथिव्यां सर्व मानवाः

*This verse of the Manusmriti reveals the fact that people of the world have sought education in Bharat for their character building and nobility of actions.*

But today we have no self-confidence, which is the cause of present misery. The only reason for this is the forgetfulness of power of self-realization.

Before we discuss ancient principles, let us remember the All-Religion Conference held in Chicago in the year 1893, where one of our *Manishis* had revealed the secret of *Bhartiy* philosophy to the whole world. In that very moment this *Yugmanishi* Swami Vivekanand impressed people from all religions.

Swami ji had lost all his belongings, including his passport. In order to avoid a severe cold, he entered into a fruit basket and got himself covered with hay. In the morning, when Swami ji was removing grass and hay above him, co-incidentally one of the organizers of the conference saw Swami ji. He instantly recognized Swami Vivekanand. Seeing his glory, he wished to talk to Swami ji. On hearing the purpose of Swami ji's coming from Bharat to Chicago (America) and the mishap of losing his belongings, the organizer took him to his house.

In the conference only five minutes were given to Swami Vivekanand for his discourse. But in the same five minutes Swami ji introduced the great *Bhartiy* philosophy in his divine voice, which captivated and hypnotized the whole audience. He also hoisted the ideological victory flag of Bharat on the whole world. The whole world only not lauded openly the sublime cultural and philosophical consciousness of Bharat, but also accepted its superiority.

*Bhartiy* thought of life does not uphold epicurean philosophy of 'Eat, drink and be merry' as its goal. It holds the collective sense of life as the goal of life and salvation as its intended purpose. *Moksh* means liberation from life and death. The belief in rebirth has created great values of life. This is stated in the *Ishophanishad* :

ईशावास्यमिदं सर्व यत्किंच जगत्यां जगत्।
तेन त्यक्तेन भुञजीथाः मा गृधः कस्यस्विद् धनम्।।

*That is, the whole creation is covered with God. After surrendering to him, enjoy whatever is left. Do not covet for other's wealth. It means that we acquire the maximum amount of wealth, so that we can serve God in a worthy manner.*

Take only as much for consumption as it is not an obstruction in your service of the society. The same thought has been clarified in the *Manusmriti* :

यावत् भ्रियेत जठरं तावत् स्वत्वं हि देहिनाम्।
अधिकं यो ऽभिमन्येत् स स्तेनो दण्डमर्हति।।

*That is, man has the right only as much as is necessary for the condition of the body. He who consumes more than this is a thief and punishable.*

Naturally if such a feeling arises, our thinking becomes socially-oriented and a pure sense of service is kindled. It automatically eradicates ego from mind and from it, a sense of collectivity and inclusivity is born. The *Vedic Rishis* proclaimed:

अग्रतः चतुरो वेदाः पृष्ठतः सशरं धनुः।
इदं ब्राह्मं इदं क्षात्रं शापादपि शरादपि।।

*One who is well versed in the four 'Vedas' and supports the bow and arrow of bravery upon his back will destroy the evil doer either with a curse or with an arrow.*

With this moral strength *Rishis* proclaimed 'कृण्वन्तम् विश्वम् आर्यम्' *(keep making the world superior)*. With this moto *Bhartiy Rishis* travelled throughout the world. They saw Bhagwan in every creature. They gave the massage of compassion, forgiveness, love and empathy to the world. It was because of this massage that Bharat got the prestige as a 'जगत गुरु' (world teacher).

सर्वे भवन्तु सुखिनः सर्वे सन्तु निरामयाः।
सर्वे भद्राणि पश्यन्तु मा कश्चिद्दुःखभाग्भवेत्।।

बृहदारण्यक उपनिषद् 1.4.14

*May all sentient beings be at peace, may no one suffer from illness, May all see what is auspicious, may no one suffer. Om peace, peace, peace. (The Brihdarnyak Upnishad 1.4.14)*

मातृवत् परदारेषु, परद्रव्येषु लोष्ठवत्।
आत्मवत्सर्वभूतेषु यः पश्यति सः पण्डितः।।

नीतिसार

*He is a wise man who sees the wives of others as his mother, the wealth of others like clod of earth and all beings as his own self. (The Nitisar)*

Such contemplations are an integral part of *Bhartiy* culture which is not to be seen anywhere else in the world literature. Gauhar Bano was brought in front of Shiv who founded the *Hindupad-Padshahi*. Shiv said to that beautiful lady, "Mother! If Shiva's mother was so beautiful Shiv would have been beautiful too". Gauhar Bano was sent back to her family with respect and dignity. This example is sufficient to understand the basic *Aatma*, the basic essence, of *Bhartiy* culture. The following *Shlok* advocates co-existence:

ॐ' सह नाववतु। सह नौ भुनक्तु।। सहवीर्य करवावहै।
तेजस्विना वधीतमस्तु माविद्विषाव है।

कृष्ण यजुर्वेद तैतिरीय उपनिषद् 2.2.2

*May God protect us both (the teacher and the student. God nourish us both. We work together with energy and vigour, our study be enlightening, not giving rise to hostility. Om Peace, Peace, Peace. (The Krishan Yajurved Tattiriya Upnishad 2.2.2)*

*Rishi* tradition of Bharat gave us the broadness and all pervasiveness of *Bhartiy* philosophy in the *Mantr* in the *Chhandoyg Upnishad* 'एकोऽहम् बहुस्यामि' *(I am one may I be many).*

त्यजेदेकं कुलस्यार्थे, कुलम् ग्रामस्यार्थे त्यजेत्।
ग्रामं जनपदस्यार्थे आत्मार्थे पृथ्वीं त्यजेत्।।

महाभारत आदिपर्व 7,37

*Sacrifice one person for the sake of the family; give up a family for the sake of a town; sacrifice of a town for the benefit of the nation; also, leave the earth for the benefit of the soul. (The Mahabharat Aadiparv 7, 37)*

This type of inclusive sentiment is the distinctive feature not to be found expressed anywhere else other than in *Bhartiy* culture and *Bhartiy* scriptures.

# Spiritual Vision of Bharat

The spiritual vision of Bharat is both expansive and generous. 'एकम् सत् विप्राः बहुधा वदन्ति।' *(truth is one, the Rishis tell this truth in various ways)*. This *Vedokti (proclamation by Ved)* gives recognition to all the sects. Hindu society recognizes thirty three categories of Gods and Goddesses (twelve Aditya, eight Vasu, eight Rudra, one Indar and one Prajapati). Anyone who believes in these thirty three categories of deities is a Hindu. However, a person who does not believe in these thirty three categories, is also a Hindu; an atheist is also a Hindu. Both worshiper of formless-*Brahm* and a worshiper of embodied *Brahm* are Hindus. This freedom in system of worshipping have given birth to several ideologies in Bharat. Many creeds like Arya Samaj, Jainism, Sikhism, Buddhism, Lingayat, Shakta, Shaiva, Vaishnav etc. are born from its cardinal spirit of inclusivity.

All religions in the world cannot be called *Dharm*. In fact, they are only sects or creeds. Unfortunately, *Dharm* and sect have been shrewdly used as synonymous. It might have sprung from evil designs of degrading Bharat. This is the reason why the noise of a new opinion called 'Secularism' is being heard all around. Sect and creed are synonymous, but *Dharm* is completely different. The Hindi translation of secularism as 'धर्म निरपेक्ष' (religion neutral) by the so-called intellectuals is grossly wrong. Creed or Sect simply imply path or paths to reach the ultimate truth and the English translation of this is 'Religion'. The literal translation of secularism should be non-sectarianism. The same translation was done in the early version of our Constitution but some politicians changed this interpretation of the Constitution because of their political interests.

# Dharm : Etymology and Scope

The word *Dharm* is derived from 'धृत धातु' *(dhrit dhatu)* by adding suffix 'मन' (mann). The meaning of 'ध'' is 'धारणा' (*dhaar-naa*), notion/ belief. That is why it is said that 'धारणात् धर्म इत्याहुः धर्मो धारयति प्रजाः' i.e. which can be imbibed is *Dharm*. Our *Rishis* proclaimed 'यतोऽभ्युदयानिः श्रेयस्सिद्धिः सः धर्मः।' *Dharm* is said to be cosmic advancement and transcendental well-being. Cosmic and metaphysical duties are considered to be *Dharm*. In Bharat, it is said that 'वेदोऽखिलोधर्ममूलम्' i.e. *Veds* are considered to be the origin of *Dharm*. As already discussed previously, it is clear that *Dharm* is not a method of worshipping.

Had *Dharm* been a method of worshipping, no king would have been as 'धार्मिक' (religious) as Ravan was. In *Bhartiy* culture and thinking Ravan was never accepted as *Dharmik*. Year after year though Ravan pleased Bhagwan Shiv through rigorous *Tapshchrya*, he was still referred to as a Demon King. In Bhagwan Shri Ram's life there is no mention of hard work or *Tapshchrya* except establishment of a Shivling at Rameshwaram. The establishment of a Shivling in Rameshwaram was also an urgent need of the hour, as Shaivism was prevalent in the Southern Bharat while Vaishnavism was dominant in the Northern Bharat. Nonetheless, Bhagwan Shri Ram is considered as *Dharmik* (religious).

During his sojourn in the forest along with his wife Sita and the younger brother Laxman, he saw the bones of innocent people killed by the demons. On seeing these bones Bhagwan Shri Ram took a pledge to destroy demon-culture. To actualize such a pledge he neither had power, army, arms nor means. During their sojourn in the forest *Rakshraj* Ravan kidnapped Sita. Bhagwan Shri Ram was in search of her and had to fulfil the pledge. For this he needed the army, the army needed weapons and both of these required great resources, which had to be collected locally. During this very period while Bhagwan Ram was engaged with this project, Ravan propagated through his detectives that Bhagwan Ram is *Vaishnav* while Ravan and his clan were *Shaivas*. Bhagwan Ram's attack on Ravan meant *Vaishnavas'* attack on *Shaivas'*. This vicious propaganda caused tension and the projects of Bhagwan Shri Ram was stalled. As a result, an emergency strategy was prepared to deal with this crisis. The result of this strategy was establishment of a Shivling in Rameshwaram. Ravan was invited to act as an *Acharya* for the establishment of the Shivling. On the basis of the available evidences, millions of tribes of the area attended this function. In front of the inhabitants of mountains, the forest dweller Bhagwan Ram clearly announced :

शिव द्रोही मम दास कहावा। सो नर सपनेहुँ मोहि न भावा।।

रामचरितमानस

*Anyone who does not believe in Bhagwan Shiv cannot be my Bhakat. Even in dreams I cannot like such a person. (The Ramcharitmanas)*

If the establishment of the Shivling was an outcome of reverence of Bhagwan Shri Ram for Bhagwan Shiv, then after Bhagwan Shri Ram's coronation such celebrations would also have taken place in Ayodhya. We do not find any evidence that after becoming a king, Bhagwan Shri Ram held any such event. Our *Rishis* have said :

सत्यं वद धर्मं चर, असतो मा सद्गमय, तमसो मा ज्योतिर्गमय ।।

तैतीरिय उपनिषद् 11.1

*Speak truth and do according to Dharm, Lead me from untruth to truth. Lead me from darkness to light. (The Tattiriya Upnishad 11.1)*

In the light of these statements, when we weigh the so-called religions of the world, we find that they are only sects and creeds, which have wrongly been termed as *Dharm*. From this one can easily conclude that *Sanatan Dharm* or Hindu *Dharm* of Bharat can only be called *Dharm* because it entirely fits in to the definition of *Dharm*.

When we consider the entire *Bhartiy* theology in the *Veds*, *Puranas*, *Upnishads*, *Manusmriti*, *Yajnvalky Smriti* and Commentaries by Medhatithi and Mitakshara, the picture we get of Hindu *Dharm* or *Sanatan Dharm* is described in six parts.

The grand edifice of Hinduism or *Sanatan Dharm* was built on the following 6 strong pillars :

1. *Varn Dharm*
2. *Ashram Dharm*
3. *Varn Ashram Dharm*
4. Attribute *Dharm*
5. Atonement *Dharm*
6. General *Dharm*

# Varn Dharm

Now a days *Bhartiy Varn System* is being attacked from all sides. From the Mughal period till the present day, the *Dharm* has systematically been distorted. This distorted version of *Dharm* has been considered as the cause of social misery. This is a lie through theeth. The truth is that this present distorted form of *Varn System* is the result of the abominable and spiteful plan of invaders, Muslims and the British. I will discuss this subject in detail later on. To understand *Varn Dharm* in a nutshell, it is enough to understand that *Varn System* was not on the basis of birth, but was based on attributes and *Karm* of any individual. According to verse 109 of the chapter I of the *Manusmriti* :

आचारः परमो धर्मः श्रुत्युक्तः स्मार्त एव च।
तस्मादस्मिन्सदायुक्तो नित्यं स्यादात्मवान द्विज।।

*The meaning of this verse is that conduct of an individual is of utmost importance. Any individual who wants to take care of his soul should always follow the rules of good conduct. Such a person is called Dwij.*

According to *Maharishi* Manu, *Brahmans*, *Kshtriys*, and *Vaishys* are *Dwijs*. The literal meaning of *Dwij* is twice born. The moment a person is born from the mother's womb, he comes under the category of *Shudr Varn* because at that time he has no knowledge and education. With knowledge, education and through his efforts whatever position he attains determines his *Varn*. By fulfilling the prescribed qualifications any person can become *Brahman*, *Kshtriy* or *Vaishy*. Thus, he becomes *Dwij* (twice born). This has been clearly mentioned in the *Manusmriti* verses 110 and 111 :

जन्मना जायते शूद्रः संस्कारात् द्विज उच्यते
वेदाभ्यास भवति विप्र ब्रह्मम जानति ब्राह्मणः

*Whoever took birth is Shudr. By sanskars one becomes Dwij, But by purification of mind and attaining Vedic knowledge by faith one become Vipr, and the one who knows Brahm (supreme soul) can only be called Brahman.*

A similar principle was recognized by Bhagwan Shri Krishn in the *Shrimad Bhagwad Gita*.

चातुर्वर्ण्यं मया सृष्टं गुणकर्मविभागशः ।
तस्य कर्तारमपि मां विद्धयकर्तारमव्ययम्।।

*According to the differentiation of attributes and actions (Karm), I have created the four Varns. Though I am the creator of Varns yet know Me to be the Non-performer.*

As I have already mentioned, Muslim invaders and power-mongrels driven by their vested interests have corrupted and destroyed the sanctity of *Dharm*. This gluttony of attempts continues till date. In various religious texts including the *Manusmriti* many verses and chapters were falsely and viciously interpolated. A serious scrutiny reveals that tone, tenor, and syntactic interpolations are not in sync with those of the original texts and are contradictory in nature.

Happiness and prosperity can be expected only with a balance between the individual and society. The purpose of this *Varn System* was to create and maintain the balance. For any system, there is a need for a code of conduct that could be fare to every unit of the society. From the *Vedic* period to the *Puranik* period, all the *Rishis* sincerely believed the society to be a manifestation of God. Because of this belief they evolved fair systems. Maharaj Manu is also one of them. However, in the course of time, certain verses were interpolated by some hypocrites, which can be understood only after a thorough scrutiny. There cannot be a healthy debate if someone has not read and understood the texts in proper context. There is abysmal ignorance. Those who have not read even a word of these texts till date are busy in cursing and abusing the *Manuwad* and Hindu scriptures because of their blindness, selfishness and ignorance.

At present, we are discussing *Varn Dharm*. So in a nutshell we can say that the *Varn System* was based on the inherent attributes and *Karm* of an individual and that individual can attain the *Varn* of his choice through his efforts and actions. The performance of the sublime duties prescribed by that *Varn* will be his *Dharm*.

# Ashram Dharm

For social order, our thinkers conceived of four *Ashrams* by dividing human life into four following stages :

1. *Brahmachary Ashram*
2. *Grihasth Ashram*
3. *Vanaprasth Ashram*
4. *Sannyas Ashram*

In short, the human beings perform their duties prescribed by the thinkers as per these four *Ashrams*. Following these four *Ashrams* is adherence to *Dharm*. History is a witness that society remained healthy and prosperous as long as humans beings followed this *Ashram Dharm*. After this system got corrupted and distortions started taking place, it affected *Bhartiy* society like cancer.

***Varnashram Dharm***

There is an interdependent relationship between *Dharm* and *Ashram* system. Therefore, *Varnashram Dharm* mandates following the rules prescribed by thinkers for strengthening the inter relationship between *Dharm* and *Ashram* system for their smooth functioning.

***Gun-Dharm***

According to the belief of the *Bhartiy* thinkers, the *Gun* (distinctive attributes) by which a person, thing or an institution is identified and due to the extinction of which such a person, things or an institution loses its individual identity is called the *Gun-Dharm*. For example the distinctive attribute or *Gun-Dharm* of fire is flammability; the distinctive attribute of water is coldness and the distinctive attribute of sugar is sweetness.

Maintaining relationships formed by birth also comes under the *Gun-Dharm*. i.e. father-*Dharm*, son-*Dharm*, mother-*Dharm*, fraternity, husband-*Dharm*, wife-*Dharm*, clan-*Dharm*, society-*Dharm*, *Rashtr-Dharm* etc. Certainly, the discharge of such assigned duties will be considered as following the *Gun-Dharm*.

**Atonement Dharm**

To err is human nature. Penance acts both as a damage controlling measure and a warning for preemption of the same mistakes in future. Through penance human conduct is refined. That's why our thinkers have framed the rules for Atonement *Dharm*. The observance of this *Dharm* is essential for social and individual morality. An individual is the basic unit of society. Individuals' fulfillment is essential for forming a creative back drop for the society. Refinement of the individual definitely gives the right direction to the society. With it our *Rashtr* can be made strong and a global leader. This is called Atonement *Dharm*.

**General Dharm**

The normal code of conduct for a human life is called General *Dharm*. It can also be called 'General Law of the Life' in English. The aim of human life is to move from darkness to light, from ignorance to knowledge. By eliminating the evil, a person assimilates the good into his life. In fact, only such a person can be called virtuous. This upward positive journey makes a person *Rishi*, *Maharishi*, Deity and finally a God. Thinkers have termed these ascending qualities as 10 signs of *Dharm*. These very qualities are given as definition of *Dharm*.

धृतिः क्षमादमोऽस्तेयं शौचमिन्द्रियनिग्रहः ।
धीर्विद्यासत्यमक्रोधो दशकं धर्म लक्षणम् ।।

मनुस्मृति 6.61

*Patience, Forgiveness, constant discrimination, non-stealing, purity, control of senses, righteous action, knowledge, truth and giving up anger - these are the ten indication of Dharm. (The Manusmriti 6.61)*

Apart from this, following attributes prescribed by other scriptures are also considered as following General *Dharm*. These attributes are truth, kindness, austerity, defecation, tolerance, sense of truth and untruth, control over mind, restraint over senses, non-violence, renunciation, self-study, contentment, being observant, good thoughts, self-realization, abstinence, thinking of God, social service, service of poor people, *(Janardan Seva)*, overcoming ego and sentiments of inclusivity etc.

Those human beings who imbibe these attributes are said to be following the General *Dharm*. Since these attributes are good for ideal human life, many scholars consider this to be '*Dharm*'. In Hindu or *Sanatan Dharm*, the qualitative journey of human life is based on these attributes. That is why these are called *Dharm*.

Apart from this, the 'पुरुषार्थ चतुष्टय' (four *Purusharthas*) prescribed for human life by sc riptures are also considered to be a part of *Dharm*. These four *Purusharthas* are *Dharm* (Righteousness), *Arth* (Wealth), *Kaam* (Desire) and *Moksh* (Salvation). Based on this, the four *Purusharth* set by our thinkers are definitely within the scope of *Dharm*. These can be explained in the follo wing way :

**Purusharth Dharm**

Hindu *Dharm* has prescribed four *Varns*, four *Ashrams* and four *Purusharthas* for human life. Just as the rules were laid down by the *Rishis* for the *Varn* and *Ashram* system, similarly the thinkers also formulated rules for the four *Purusharthas*. These four *Purusharthas* are *Dharm*, *Arth*, *Kaam*, *Moksh*.

Where the *Dharm* is for the organization of the society, the *Ashram* system and four *Purusharthas* are meant to organize human life. The society can be organized only by organizing individuals, as individuals are the basic unit of the society. Such was the opinion of our thinkers. Therefore, scriptures were composed considering human beings, as the basic unit of society, which forms the center of all *Dharmik* systems.

After getting human life, it is the duty of every human being to make efforts for *Dharmcharan* (to follow the path of *Dharm*) so that he can act righteously. In this way he behaves virtuously and with purity for the rest of his life. A person treading the wrong path definitely becomes a burden on the earth. Excellent and refined attributes are righteousness.

Earning wealth, with a *Dharmik* attitude also falls in the category of *Purusharth*. By following these prescribed rules for earning wealth a person definitely performs the *Purusharth Dharm*.

The culture of Bharat is the culture of *Satyam, Shivam, Sundaram (truth, welfare, beauty)*. That is why our thinkers have placed *Kaam* (desires) in the category of *Purusharth*. Naturally, its discharge also falls under *Purusharth Dharm*.

*Moksh*, the attainment of salvation is the ultimate goal of human life. The person who desires salvation remains careful and attentive towards performance of duties. This is his *Purusharth*. The last step for *Moksh* is the *Dharm*-based earning of wealth and fulfillment of desires. It is due to the acceptance of these *Purusharthas* that *Bhartiy* culture is considered as the culture of *Satyam, Shivam, Sundaram*. Performing these *Purusharthas* will be called *Purusharth Dharm*.

After seeing *Dharm* in this perspective, the difference between *Dharm* and Sect can easily be understood. Goswami Tulsidas has revealed the prevalence of *Dharm* in the following way. :

परहित सरिस धर्म नहि भाई।
परपीड़ा सम नहि अधमाई।। रामचरितमानस

*The biggest Dharm is the interest of the other and the biggest wrongdoing is to cause trouble to the other. (Ramcharitmanas)*

This concept of *Dharm* is naturally acceptable to the people of Bharat. The *Manusmriti* has clearly stated that :

धर्म एव हतो हन्ति धर्मो रक्षति रक्षितः।
तस्माद्धर्मो न हन्तव्यो मा नोधर्मोहतोऽवधीत।। 8,15

*Those who kill Dharm will be destroyed completely by it. Those who protect Dharm will be protected by it. Therefore, never destroy Dharm so that Dharm can protect us.(8,15)*

Due to this lofty and superior comprehensive concept of *Dharm*, the *Manusmriti* says :

एक एव सुहृद्धर्मो निधने ऽप्यनुयाति यः।
शरीरेण समं नाशं, सर्वमन्यद्धि गच्छति।। 8,17

*That is, Dharm is the only friend that walks along after death, otherwise everything is destroyed in this world after the body is destroyed. (8, 17)*

A human being sans *Dharm* is an animal. By saying this *Bhartiy* scriptures proved the significance and essentiality of *Dharm* in human life. It is the *Dharm* that distinguish humans from animals despite there being similarities between them.

आहार–निद्रा–भय–मैथुनं च। सामान्ययेतद् पशुभिः नराणाम्।।
धर्मो हितेषाम्ऽधिको विशेषो। धर्मेणहीनः पशुभिः समानाः।।

भर्तृहरि

*Food, sleep, fear and mating, these acts of humans are similar to animals. Of them (humans), Dharm is the only special thing, without Dharm humans are also animals. (Bharathari)*

# References from Scriptures

## Vedic Literature

The *Veds* are considered to be the point of genesis of Hindu or *Sanatan Dharm*, though some scholars also believe that *Sanatan Dharm* existed even before the *Veds*. However, with the manifestation of the *Veds*, it became easier to present the *Sanatan Dharm* in a comprehensive form. The *Rigved* has been called the original *Ved*. The remaining three *Veds* namely the *Yajurved, Samved* and *Atharvved* emerged from the *Rigved*. However, according to the *Manusmriti*, the three *Veds* namely the *Rigved*, *Yajurved* and *Samved* were first manifested by *Brahm*. The *Atharvved* came into existence by separating the *Mantrs* and ritualistic procedures from these *Veds*. *Maharishi* Angira is said to be editor or redactor of the *Atharvved*. It is said that *Maharishi* Angira was an expert of ritualistic procedures. It is also believed that Brihaspati, the Guru of the Gods was himself Angira *Rishi*.

According to some scholars, the *Veds* are basically similar. The *Veds* have been divided into three parts on the basis of contents. Scholars are also of the opinion that at the time of doomsday God assigned the responsibility of protection of the three *Veds* to *Agni*, *Vayu* and *Surya* respectively and asked them to manifest these while creating the world again.

Astu! it is clear that because of the *Veds*, a sublime and organized form of *Sanatan Dharm* was put before the world, which is as much relevant in the present times as it was at the time of its genesis. As per *Veds*, it is not any religion or sect but human existence is paramount. There is no question of caste or sub-caste. A profound spirit of coexistence is encoded in the *Veds*. That's why 'वेदोऽखिलो धर्ममूलम्, ऋषियो मन्त्रदृष्टारः' *(that is, the root of Dharm is the Veds and the Rishis are the envisioners of the Mantrs.*

The following hymns of the *Rigved* is a unique example of this spirit of coexistence and togetherness :

सं गच्छध्वं, सं वदध्वं सं वो मनांसि जानताम्।
देवा भागं यथा पूर्वे संजनाना उपासते। 10.191,2

*Let all move together, talk together in unison. Together, let all acquire knowledge in the same way as Gods did in the past. (10.191,2)*

The following example of social harmony is also remarkable :

समानो मन्त्रः समितिः समानी समानं मनः सह चित्तमेषाम्।
समानं मन्त्रभि मन्त्रये वः समानेन वो हविषा विधेम।। 10,191,3

*May Mantrs of prayer of all men be the same, may the committee of all citizens be the same, everyone's mind and heart be the same. I invoke you with the same Mantr and I make you to offer same type of sacrificial offerings, and I make you cultured. (10,191,3)*

समानी व आकूतिः समाना हृदयानि वः।
समानवस्तु वो मनो यथा वः सुसहासति।। 10,191,4

*May your determinations be the same, your minds united, and your hearts the same. so that you may be beautifully organized together. (10,191,4)*

The *Atharvved* likewise also gives key to societal and organizational behaviour. The conception of harmony and egalitarianism is the fundamental doctrinal commitment of the *Veds*. The seven verses mentioned in the 30 hymn of the 3 Kand of the *Atharvved* are exclusively devoted to this spirit of togetherness:

सहृदयं सांमनस्ययं विद्वेषं कृणोमि वः।
अन्योऽन्यमभिहर्यत वत्सं जातिमिवाध्न्या।।

*By removing the animosity from all of you, I spread the feeling of kindness and gentleness. Love each other as a cow loves its calf. Love one another.*

अनुव्रतः पितुः पुत्रो माता भवति समनाः।
जाया पत्ये मधुमतीं वाचम् वदतु शान्तिवाम्।।

*The son should have commitment to father and carry out mother's commands. Wife should talk to husband peacefully and sweetly.*

मा भ्राता भ्रातरं द्विक्षन् मा स्वसारयुतस्वसा।
सम चः सव्रता भूत्वा वदत भद्रया।।

*Brothers and sisters should not envy each other. All should speak in kind words with positive intentions and equal goodness.*

येन देवा न वियन्ति नो च विद्विषते मिथः।
तत्कृण्मो ब्रह्म वो गृहे संज्ञानं पुरुषेभ्यः।।

*Under the influence of which the Deities do not think otherwise and do not envy each other, likewise we recommend this Samta Mantr for the people of house.*

ज्यायस्वन्तश्चित्तिनो मा वि योष्ट संराध्यन्तः साधुराश्चरन्तः।
अन्यो अन्यस्मै वल्गु वदन्त एत सध्रीचीनान्वः संमनसस्कृणोमि।।

*Do not discriminate, while performing the same action with the same intention, taking care of the elder and younger. Speak in a mellow and sweet voice to each other. Come on, O human beings! I create you as capable of performing actions in the same manner.*

समानी प्रपा सह वोऽन्नभागः सामने योक्त्रे सह वो युनज्मि।
सम्यञ्चोऽग्निं सपर्यतारा नाभिमिवाभितः।।

*All of you drink water from the same source, share the same food. I bind you in the same string as the spokes of the wheel are connected to a center. In the same way, you worship Agni (Fire) with the same purpose.*

सध्रीचीनान्वः संमनसकृणोम्येकश्नुष्टीन्त्संवननेन सर्वान्।
देवा इवामृतं रक्षमाणाः सायंप्रातः सौमनसो वो अस्तु।।

*I envision you to engage in one task, have one opinion, make you eat the same food. Through this hypnotic act, I exert all of you like Indar and other Deities in the heaven, who make people immortal and non-aging and who protect the nectar in heaven. Let your heart be beautiful all through the day.*

The above *Mantrs* of the *Atharvved* give the same sense of affinity without any disparity. Social welfare is ensured with the spirit of coexistence between human beings. Social welfare includes individual's welfare. This is the essence of above cited *Mantrs*.

*Vedic* belief regarding the *Dharm* is not based merely on birth of a person but is based on distinctive attributes and *Karm*.

When the *Veds* do not differentiate between human species, how can they differentiate between their progeny? That is why in the *Vedic* period, all people had the same right to perform *Yagy* and other religious rituals.

In the following verse of the *Rigved*, the *Rishis* asked people from all five categories to perform *Yagy* and share its fruits.

तदद्य वाचः प्रथमं मसीय येनासुराँ अभि देवा असाम।
ऊर्जाद उत यज्ञियासः पञ्च जना मम होत्रं जुषध्वम्।।

10,53,4

*Today I am pronouncing that main speech, with which Gods could defeat Demons. O five people! (Brahmans, Kshtriys, Vaishys, Shudrs and Nishads) who share the food and have the authority of the Yagy, you should consume the fruits of my havan. It means you perform Yagy as directed by me. (10,53,4)*

Similar sentiments are found in other verses of the *Rigved*. The interpreters have given a comprehensive explanation of 'पंञचजना मम होत्रं जुषध्वम्' (*five Varns eat the fruit*). According to them - 'चत्वारो वर्णानिषादः पञ्चम् इत्यौपमन्यवः'.

According to this 'पंञचजना ' (*five Varns*) means *Brahman*, *Kshtriy*, *Vaishy*, *Shudr* and *Nishad*. Many scholars have not accepted the fifth *Varn Nishad* in addition to the four *Varns*. In this context, the name of Swami Dayanand Saraswati, the founder of Arya Samaj is referred. In commentaries of the *Rigved* Swami Dayanand Saraswati ji has taken the meaning of *Nishad* as *Avarn*. According to him, God preaches that the four *Savarns* and the fifth *Avarn* should also perform the rituals of the *Yagy* as prescribed by him. Being the performers of the *Yagy* they share its fruits. The opinion of the scholars in this regard is also that the word *'Savarn'* will be used for all the people coming under the *Varn System* and the rest will be called *Avarn*. The words *'Savarn'* and *'Harijans'* as prevailing today are neither supported by scriptures nor are appropriate. It seems that such a distinction was propagated only after fourth century. Over the time, this distinction was viciously made to culminate in untouchability. Otherwise, it has been clear that who ever are included in the *Varn System* are *Savarn*. The one who does not come under this system is *Avarn*. *Brahmans*, *Kshtriys*, *Vaishys*, *Shudrs* all are *Savarns* not on the basis of birth but on the basis of their distinctive attributes and *Karm*.

Similar views are also seen in the *Apastamb-Dharmsutr*: 'आर्याधिष्ठिता वा शूद्राः संस्कर्त्तारः स्युः।' (*according to this Sutr, the Shudrs along with others are also entitled to food under the care of the Aryans - 2.1.18*)

A clear reference is available in the *Apastamb-Dharmsutr* regarding change from one *Varn* to another :

धर्म चर्या जघन्यों वर्णः पूर्व वर्णमापद्यते जाति परिवृत्तौ।
अधर्म चर्याया पूर्वो जघन्यं वर्णमापद्यते जातिपरिवृत्तौ।।

10,11

*The meaning is that by practicing Dharm an individual can assume in different castes/Varn. Thus Kshtriys can become Vaishys, Shudrs and Brahmans. Vaishys can become as Kshtriys, Brahmans and Shudrs. Shudrs can become Brahmans. This is possible only by following or not following Dharm. (10, 11)*

In fact, the orthodox system of today did not prevail in the ancient times. The truth is that the view of the ancient Indian scholars was very broad and expansive. They did not even distinguish between their kins and strangers. This discrimination is the result of parochial thinking. For those who have a generous heart, the whole world is a family.

अयं निजः परो वेति गणना लघु चेतसाम्।
उदारचरितानां तु वसुधैव कुटुम्बकम्।

Also how meaningful and relevant is this proclamation of the ancient *Rishis* in the present times!

अभयं नः करत्यन्तरिक्षमभयं द्यावापृथिवी उभे इमे।।
अभयं नः पश्चादभयं पुरस्ताधराद्भयं नो अस्तु।।
अभयं मित्रादभयन्मित्रादभयं ज्ञातादभयं परोक्षात्।
अभयं नक्तमभयं दिवा नः सर्वा आशा मम मित्रं भवन्तु।।

अथर्ववेद 19–2,6,5–6

*No one should be afraid of society, space, heaven, earth. Let everybody be safe back and forth, upside-down. Let everybody be fearless of friend or enemy, known or unknown, day and night. Nobody should be affected or influenced from any direction. This is the auspicious prayer for everyone. (Atharvved 19-2,6,5-6)*

I question the so-called self-proclaimed scholars who advocated depriving *Shudrs* and women of the opportunity to study the *Veds* by saying 'ना स्त्री शूद्रो वेदमधीयताम्'. In this country Kvas, Ailush, Mahidas, Aitareya, Raikva *Rishi*, Aushij, Kakshivan, Vatsa Kanvayan, Sudakshina, Kshaimi, Satyakam, Jaba and female *Rishis* like Apala, Atreyi, Vishwavara, Kosha, Kakshvati, Lopamudra, Sikta Nivavari and Romasha and other great theologians and knower's of *Aprvidya* belonged to different *Varns* by birth. If this is so, how is it possible that females and *Shudr Varns* were not allowed to study the *Veds*. Many of the above referred *Rishis* and *Rishikas* were enunciators of several *Mantrs* found in the *Veds*. This partisan declaration of the so-called scholars is socially fatal. How can persons who commit such crimes be well-wishers of *Sanatan Samaj* or Hindu society? In my view, they are not well-wishers.

The 18th *Sukt* of the 10th *Mundal* of the *Rigved* is called *Purushsukt*. This hymn is very famous and popular in the priestly class. In every ritual, whether there is *Yagy*, *Vivah*, *Shradh* or any other rites, the verses of this *Sukt* are recited by the priests. Many scholars consider this entire hymn to be a later composition. This *Sukt* of the *Rigved*, which has 16 *Mantrs*, is also in the *Yajurved* and the *Atharvved*, but there is a difference in the order of their appearance. Among them, a *Mantr* which mentions the origin of the four *Varns* is criticized by many a scholars. The *Mantrs* is :

ब्राह्मणोऽस्य मुखमासीदबाहू राजन्यः कृतः।
ऊरू तदस्य यद्वैश्यः पद्भ्यां शूद्रो अजायत।

*Brahmans from the face, Kshtriys from the arms, Vaishy from the thighs and Shudrs are born from feet.*

Some scholars say that 11th and 12th verse in *Purushsukt* were lateral interpretations. According to these scholars, the style and language of these *Mantrs* are completely different from that of the rest. These two verses also seem to disturb the continuity of thought between the 10th verse and the 13th verse.

Dr. Ambedkar, while severely attacking the *Purushsukt*, has called the *Varn System* biased. His objection is mainly on the 'पदभ्यां शूद्रो अजायत'. According to this *Sukt*, *Shudr* originated from the feet. In fact, if we think impartially and without prejudices, it would be clear that some scholars have misinterpreted these verses from *Purushsukt*. If we keep the basic spirit of the entire *Mantrs* in our mind then the true meaning will not be lost. If we miss even a little then we will never be able to understand the holistic point-of-view. For example, if one has to go to Mumbai and gets on to the train going to Calcutta, then the journey of the train will continue but the goal will not be achieved. In the same way, if any scripture is to be interpreted, then the holistic point-of-view contained in its totality has to be kept in view. If the verse from the *Purushsukt* are to be interpreted in the back drop of the extent to which the concept of co-existence is described in the *Vedic* literature by the *Rishis*, one will not be led into any wrong interpretation. This verse will not be put on scaffold. Renowned scholars like Satvalekar and the Kale have interpreted this verse in this way only.

According to Satvalekar and Kale, the *Adipurush* in this verse is none other than *Sanatan* Samaj and similarly *Yagy* is nothing but the restoration of the social structure. This social structure should be seen as a complete body. As no part of the body can be graded or separated from the whole, likewise no *Varn* is high or low. If any part is separated, body would get impair and would collapse. In the same way, without an integrated *Varn System* this vast body of the social structure will also be infected and disabled.

The meaning of verse from *Purushsukt* was not correctly understood because the interpreters did not pay attention to proper meaning of *Varn System*.

Dr. Ambedkar gave importance to scholars who interpreted the term 'पद्भ्याम्' by considering it as fourth inflection i.e. dative case. In fact the word 'पद्भ्याम्' *Padbhyam* is the same in third inflection i.e. instrumentative case, fourth inflection i.e. dative case and the fifth inflection i.e. ablative case, but their meanings will be differently interpreted in different inflection. If 'पद्भ्याम्' is interpreted as per the fourth inflection, then it will mean 'to his feet', while considering the fifth inflection, it will mean 'from his feet'. The meaning of this *Shlok* has also been interpreted by some people as fourth inflection. Therefore the meaning of the *Shlok* has become distorted and disputed. If it is interpreted correctly, then there will be no reason for dispute.

*Viraat Purush* created a *Yagy Purush* whose body parts are head, hands, thighs and feet. These four body parts are four parts of *Varn System* respectively. People with positive vision understand that four parts of the *Varn System* form the four important segments of the society. The nature of the interpersonal relationships among these four segments will be of that of interdependence. Brain controls the whole body. If someone hurts the head, arms, stomach, legs or any other part of the body, the hands immediately come forward to save the rest of the body parts. In this process they might injure themselves. Food is chewed inside the mouth. After making it digestible mouth sends it to the stomach. The stomach gives the necessary nutrition to the entire body as per the requirement. Mouth and stomach do not keep anything for themselves but distribute the food evenly throughout the body as needed. If there is an injury or a thorn pricks feet, the brain immediately feels the pain and expresses it by shedding tears from the eyes and takes action by thinking of remedies. Accordingly, the hands become active to remove the pain in the foot. In this way all the parts of the body have a wonderful and interdependent relationship. Similarly, there should be a cohesive and conducive relationship between the four *Varns* for the betterment of the society. This is reason why the Creator imagined the society as a composite whole and a complete body in the *Purushsukt*. If this analogy is correctly understood, there will be no question of any kind of criticism. Concrete proof of this is given in the verses before and after the *Purushsukt*.

After the *Rigved*, when we look at the other three *Veds*, *Purans* and *Upanishads*, we find a strong support for this hypothesis.

All the 16 verses of the *Purushsukt* are quoted with the intention that the readers can understand the fundamental spirit of the *Veds* and through this understanding can effectively counter the alleged criticisms and unrestrained propaganda against the Hindu scriptures :

सहस्त्रशीर्षा पुरुषः सहस्त्राक्षः सहस्त्रपात् ।
स भूमिं विश्वतो वृत्वाऽत्यतिष्ठद् दशाङ्गुलम् ।।

*The Perfect Being has thousand (unlimited) heads, thousand (unlimited) eyes, and thousand (unlimited) feet. Having pervaded the whole earth (manifest universe), he remains ten fingers above (i.e. He is limitless). (1)*

पुरुष एवेदं सर्व यद्भूतम् यच्च भव्यम्।
उतामृतत्वस्येशानो यदत्रेनातिरोहति।।

*The present, past, and future (the three periods) are the Perfect Being. And (He) is the Bhagwan of immortality, and all that grows and develops with food. (2)*

एतावानस्य महिमातो ज्याजाँश्च पूरुषः।
पादोऽस्य विश्वा भूतानि त्रिपादस्यामृतं दिवि।।

*Such is His greatness, and the Perfect Being is greater than this. The manifest universe is only his one fourth (a quarter); His three-fourth, which is immortal (unmanifest), is in the heavens.(3)*

त्रिपादूर्ध्व उदैत पुरुषः पादोऽस्येहाभवत् पुनः।
ततोविष्वङ् व्यक्रामत् साशनानशने अभि।।

*With three-fourth, the Perfect Being rose upwards; one-fourth of Him again remained here. Then He spread on all sides over what eats (living beings- humans, animals, plants), and what does not eat (the inanimate). (4)*

तस्माद् विराटजायत् विराजो अधि पूरुषः।
सजातो अत्यरिच्यत् पश्चाद्भूमिमथो पुरः।।

*From Him, Viraat was born, from Viraat the Purush (in the form of the individual jiva). As soon as he was born, (he) separated himself (from Viraat in the forms of gods, human beings, animals etc.), then (he created) the earth, and then the body. (5)*

यद् पुरुषेण हविषा देवा यज्ञमतन्वत।
वसन्तो अस्यासी दाज्यं ग्रीष्म इध्मः शरद्धविः।।

*When the gods performed a sacrifice with the Perfect Being as the oblation, the Spring was its ghee (butter), the Summer its fuel, and the Autumn its oblation. (6)*

तंः यज्ञं बर्हिषि प्रौक्षन् पुरुषं जात्मग्रतः।
तेन देवा अयजन्त साध्या ऋषयश्च ये।।

*They besprinkled the first-born Purush, as to be sacrificed, on the sacred grass. Further, the gods the 'Sadhyas', and the seers that are, all sacrificed. (7)*

तस्माद्यज्ञात सर्वहुतः सम्भुतं पृषदाज्यम्।

पशून् ताँश्चके वायव्यानारण्या ग्राम्याश्च ये।।

*From that Sarvahut (wherein all/everything is sacrificed) sacrifice, the butter mixed with curd was gathered. He created these creatures of the air (birds), of the forests (wild animals), and those of the villages (domesticated animals). (8)*

तस्माद्यज्ञात् सर्वहुत ऋचः सामानि जज्ञिरे।
छन्दासि जज्ञिरे तस्माद्यजुस्तस्मादजायत्।।

*From that Sarvahut sacrifice, were born the Rca, and the Samans, from that were born the metres, from that were born the Yajus (ritualistic formulae). (9)*

तस्मादश्वा अजायन्त ये के चोभयादतः।
गावो हा जज्ञिरे तस्मात् तस्माज्जाता अजावयः।।

*From that, horses were born, and all those that have two rows of teeth, from that, cows (cattle) were born, from that, goats and sheep were born.* (10)

यत्पुरुषं व्यदधुः कवितधा व्यकल्पयन्।
मुखं किमस्य कौ बाहु का ऊरु पादा उच्चते।।

*When they divided the Perfect Being, into how many parts did they make? What was His mouth? What was His arms? What were the two thighs, and what were said to be his two feet?*(11)

According to the commentator Sayan, 'Bydadhu' means 'संकल्पेन उत्पादितवन्तः' (born out of determination) that is, *Purush* originated from *Sankalp* by the Gods.

ब्राह्मणोस्य मुखमासीद्बाहू राजन्यः कृतः।
उरु तदस्य यद्वैश्यः पदभ्यां शूद्रो अजायत।।

*The 'Brahman' was his mouth, the 'kshatriy' was made of His two arms, then His two thighs became the 'Vaishy', from His feet the 'Shudr' was born. (12)*

The above *Mantrs* i.e. number 10th, 11th and 12th are considered by some scholars as interpolated later on. This fact has already been discussed earlier. If the interpretation of these three verses is done correctly and in a holistic way, then there would be no objection in accepting them.

चन्द्रमा मनसो जातश्चक्षोः सूर्यो अजायत।
मुखादिन्द्रश्चाग्निश्च प्राणाद्वायुरजायत।।

*The Moon was born from (His) mind, from (His) eyes the Sun was born. From (His) mouth Indra and Agni (were) born, from (His) breath Wind was born. (13)*

नाभ्या आसीदन्तरिक्षं शीर्ष्णोद्यौः समवर्तत्।
पद्भ्यां भूमिर्दिशः श्रोत्रात्तथा लोकां अकल्पयन्।।

*From (His) navel was (produced) the middle-region, from (His) head was evolved the sky, from (His) two feet the earth, and the quarters (dik-directions) from (His) ears. Thus, they made the world. (14)*

सप्तास्यासन् परिधयस्त्रिः सप्त समि धः कृताः।
देवा यद्यज्ञं तत्वाना अबध्नन् पुरूषं पशुम्।।

*Seven were His enclosing (sticks), twenty one were made (his) fuel, when the Gods performing the sacrifice bound the Perfect Being as the Purush-animal. (15)*

A few scholars associate the practice in *Yagy* to *Purush* as an enchained animal. They believe that instead of the animal, *Purush* was used as offering and the *Yagy* was performed. I believe that if the meaning of 'अबध्नन्' (Abadhanan) is taken to be 'to bind', then all the sixteen verses become relevant and there will be no encroachment upon both language and grammar. If we think imposing these sixteenth *Mantrs* on the realm of feelings then we get a very gigantic picture of society, which is full of renunciation and sacrifice. This very thought is expressed in the following verse of Adi Parv of the *Mahabharat*.

त्यजदेकं कुलस्यार्थे, कुलम् ग्रामस्यार्थे त्वजेद्।
ग्रामम् जनपदस्यार्थे, आत्मार्थे पृथ्वी त्यजेत्।।

*For the sake of the family, sacrifice the individual; for the sake of village, sacrifice family; for the sake of district, sacrifice village and for the sake of your own soul, leave the earth.*

In the *Veds* also there are many quotations along with 'तेन त्यक्तेन भुञजीथा', which proclaim the sublime tradition of sacrificing oneself for the sake of all. In the above *Sukts*, *Purush* has been shown as tied to be used as an offering in the *Yagy*. This is based on the feeling of 'तेन त्यक्तेन भुञजीथा'. In poetry many idioms, proverbs and euphemisms are used. Literal meaning should not be drawn. Literal meaning will kill the noble feelings. It will be unfortunte. Similar meanings were taken from *Dharmshastrs* either because of lack of proper knowledge or because of ignorance. Due to these wrong meanings animosity has increased and is still increasing by the day. It is imperative that the scholars who consider *Rashtr* as supreme should collectively untangle the web of these misconceptions so that the best and timeless truth can be put forward before the world.

यज्ञेन यज्ञमयजन्त देवास्तानि धर्माणि प्रथमान्यासन।
ते ह नाकं महिमानः सचन्त यत्र पूर्वे सध्याः सन्ति देवाः।।

*The deities worshiped the Yagy God through this Yagy Purush. These form were the primary Dharm, where worshipers and deities reside. These glorified sacrificers started living in the same heaven.*

If we symptomatically analyze the last verse, then the overall intention of the scriptures become clear. The social welfare is the sacrament of *Yagy* and the *Yagykartas* who, to attain heaven, perform it. By attaining the ultimate goal of life, i.e. *Moksh*, they are liberated from the cycle of birth and death and they are admired and revered. Similar views have been expressed by many commentators. I also believe that holistic way is the foundation of the *Vedic* discourse. Only those who understand the tenacity of this foundation can realize the inner beauty and harmony of the structure called society. Otherwise, the story of the elephant and four blind men becomes true. The four blind men described the elephant by the way they touched its different body parts. The one who touched the elephant's feet said it was like a pillar, the one who held the tail said it was like a broom, the one who touched the ear said it was like a fan. Scriptural interpretations will meet the same fate. Similarly, the real position of those who have a thick veil of confusion and ambiguity on their inner eyes will act like four blind men.

Actually, the *Manusmriti*, the *Yajnvalky Smriti*, the *Mahabharat*, the *Vishnu Puran*, the *Matsya Puran*, the *Vayu Puran*, the *Bhavishy Puran*, the *Shrimad Bhagwad Gita*, the *Vajrasuchyopanishad*, etc. all make it clear that the four *Varns* originated from the body parts of that *Viraat Purush*. Yet none of the four *Varns* is analogous to its organs. The body of society has four parts, four *Varns*. This is the meaning of the 12th *Mantr* of the *Purushsukt* and other *Mantrs*. It is a figurative expression, which clearly means that there are four parts of the great body of the society i.e. *Mukh-Brahman* to spread knowledge; strong arms-*Kshtriy* to protect everyone; farmer and merchant-*Vaishy* to feed everyone and the *Shudr*-workers to keep this whole body moving. This is the wonderful concept of raising the body of the vast society on the path of healthy and peaceful progress. Those who find fault in this are either delusional or intellectually bankrupt. It should have been enough for them to understand the third *Mantr* of *Purushsukt*. It says 'पादोऽस्य विश्वा भूतानि' i.e. the feet of that *Viraat Purush* are all the creatures of the world. Such a frivolous thought as '*Shudr* for feet' can only be the result of a morbid mind-set. Our great *Rishi* tradition of chanting the *Mantr* 'सर्वे भवन्तु सुखिनः' rules out all unhealthy divisive thinking. There is also a sufficient basis for people to understand that the birth of the earth has also been from the feet. Ganga is also said to have originated from the feet of *Brahma* who adorns the head of Bhagwan Shiv.

Astu! Only incisive study of the scriptures can eliminate the ignorance of such people. Hence, it is essential that one should be can did, truthful and sincere.

Another example that can go with the above intention enshrined in the *Veds* is this quote from Panini's teaching :

छन्दः पादौ तु वेदस्य, हस्तौ कल्पोऽथ पठ्यते।
ज्योतिषामयनं चक्षुर्निरुक्तं श्रोत्र मुच्यते।।
शिक्षा घ्राणं तु वेदस्य मुखं व्याकरणं स्मृतम्।
तस्मात् साङ्गधीत्यैव ब्रह्मलोके महीयते।।

*In this the Veds were also conceived as a body and the creator said that Veds are the verses of God, scriptures are feet, kalp is his hands, astrology is his eyes, nirukt is his ears, education is his nose and grammar is his mouth. The perfect attainment of such a God (body) makes one worthy of the Brahmlok.*

It is also clear from this that the legs of the body are not a symbol of inferiority but an important part of the whole body. Every part has equal importance and indispensability. The body cannot be imagined by separating various organs. Similarly, like the human body, these four parts of the body form a colossal 'unbroken' complete body by connecting to each other. Hindu or *Sanatan* society cannot survive by separating these *Varn*-like parts.

# References from Purans

*Purans* have an important place in *Bhartiy* scriptural literature. The *Vedic* concept in relation to the *Dharm* emerges from the perusal of the eighteen *Purans*. It is not possible to describe all in detail but it is certainly indispensable to cite a few references from some texts for proper rendering of the subject.

Revolutionary ideas can be found in the *Varah Puran*. In the 114 chapter of the *Varah Puran*, while talking to the *Prithvi*, Narayan himself says that leaving aside thousand of *Rishis*. I worships that *Shudr* who performs *Dharmik* deeds :

अथ शूद्रस्य वक्ष्यामि कर्माणि शृणु माधवि।
यानि कर्माणि वक्ष्यामि शूद्रो मह्यं व्यवस्थितः।।

Narayan described the deeds :

देशकालावजन्तौ वर्जितौ तमसो रजः
निरहङ्कारशुद्धान्तौ अतिथेर्मन्त्रवत्सलौ।।
श्रद्दधानौ विपूतान्तौ लोभमोहविवर्जितौ।
नमस्कारप्रियौ नित्यं मम चिन्ताव्यवस्थितौ।।

*One who is unaware of the spatio-temporal, is devoid of Rajogun, Tamoguna and is egoless; who is devoted to Mantrs and guests, who is faithful and has pious character, is devoid of greed and delusion; who has unflinching devotion towards me remains in my concern. I always take care of such people.*

शूद्रकर्माणि मे देवि य एवं स समाचरेत्।
त्यक्त्वा ऋषिसस्त्राणि शूद्र मेव भजाम्यहम्।।

*O Earth! leaving thousands of Rishis, I worship that Shudr who performs the actions described in this way.*

In relation to *Shudrs*, it is clear from the above-mentioned references of the *Varah Puran* that the *Puranic* views regarding the *Varn System* are the same as expressed in the *Vedic* literature, the *Shrimad Bhagwad Gita* and the *Mahabharat* etc. It is evident from these references that *Shudr* was a person neither by birth nor he was looked down upon in the society. In the *Varah Puran* itself, the detailed method of initiation for the four *Varns* is given in chapter 126 and 127. In these there are also laws for the initiation of *Shudrs*, which supports the above statement. There are many anecdotes in the *Purans*, in which a person born in any one *Varn*, because of his conduct, learning and devotion to God, can be considered superior even to *Brahmans*.

The study of the *Veds* by huntsman; superiority of the *Vaishnav-Chandal*; the *Shudr* protagonist of the *Padmanabh Dwadashi*; description of the heroine in *Varah Puran*, Shudreshwar Mahadev, the saga of the *Shudr* Paijavan, the completion of *Yagy* performance by the *Shudr* with firm mind and the contribution of the *Shudrs* to the Tirupati temple etc. are described in the *Skandh Puran*. These references clearly ridicule prevailing caste distinctions.

Similar views have also been expressed in the *Varah Puran* regarding the *Varn System*. Regarding the flexibility of *Varn*, the *Puran* says :

एभिस्तु कर्मभिर्देव शुभैराचरितैस्तथा।
शूद्रो ब्राह्मणतां गच्छेद् वैश्यः क्षत्रियतां व्रजेत।।

*O Goddess! by doing these good deeds, a Shudr becomes a Brahman and a Vaishy becomes a Kshtriy.*

एतैः कर्मफलैर्देवि न्यूनजातिकुलोद्‌भवः।
शूद्रोऽप्यागमसम्पन्नो द्विजो भवति संस्कृतः।।

*O Goddess! because of these actions and knowledge of Veds even a Shudr becomes a Dwij.*

The intention is that a person could change his *Varn* if he was virtuous, chaste and cultured. This clearly shows that *Varn System* was based not on birth but on the attainment of a particular *Varn*-attributes by a person. Clarifying this, the *Puran* further says that :

न योनिर्नापि संस्कारो न श्रुतिर्न च संततिः।
कारणानि द्विजत्वस्य वृत्तमेव तु कारणम्।।

The reason for *Brahmantav* is neither birth, nor social conditioning, nor knowledge of the *Veds*, nor being a progeny. Virtuous conduct is the only reason for *Brahmantav*. This has been further clarified in the next verse. *Purankar* has said that in this world, one is considered a *Brahman* only by conduct.

If a *Shudr* is virtuous, he will attain *Brahmantav*. Similarly, a *Brahman* without a virtuous conduct is a *Shudr*.

सूर्वोऽयं ब्राह्मणों लोके वृत्तेन तु विधीयते।
वृत्ते स्थितश्च शूद्रोऽपि ब्राह्मणत्वं च गच्छति।
ब्राह्मणश्च भवेच्छूद्रोऽत्येवंतृत्तश्च मे मतः।।

Similarly, in other *Purans* also, many verses and narratives with regard to *Varn System* address it as not being based on birth, but on virtue and *Karm*. Most of the 18 *Purans* propound this principle. After looking at the *Purans*, we can say that in ancient *Bhartiy* philosophy, the Antyajas, *Shudrs* and women were given equal rights and respect. Knowledge, devotion and spiritual practices were important and caste or gender were insignificant and meaningless.

# Bhagwat Puran

The *Bhagwat Puran* has special importance in Hindu society. Though every *Purans* has its own speciality yet *Bhagwat Puran* has a distinctive significance in the society. Due to its devotional aspect, language and sentiment, of all Purans this *Puran* is very popular. The 6th chapter of *Shrimad Bhagwat Mahatmya* of the *Uttar Khand* of the *Padm Puran*, describes the importance of this *Puran* :

इति च परम गुहयं सर्वसिद्धांतसिद्धिं।
सपदि निगदितं ते शास्त्रपुञ्जजं विलोक्य।।
जगति शुककथातो निर्मलं नास्ति किंचित।
पिब परसुखहेतोर्द्वादशस्कन्धसारम्।।

*It means that there is nothing more pious in this world than the Bhagwat Katha emanating from the mouth of Mahamuni Shuka and which is true according to all parameters of all the scriptures. It is to be preserved and is accepted in all the principles. Therefore, to attain the ultimate bliss, drink this essence of the twelve Skandhs.*

In this way, the best amongst the scriptures, this book is the authentic text of *Bhartiy* culture, contemplation and philosophy. The verses of this book related to *Sanatan Dharm* are noteworthy. In this regard, see the following verse of the 16th chapter of the 5th *Skandh* :

न जन्म नूनं महतो न सौभगं। न वाङ् न बुद्धिर्नाकृतिस्तोषहेतुः।।
तैर्यद्विसृष्टानपि नो वनौकस– श्चकार सख्ये बत लक्ष्मणाग्रजः।।

*Bhagwan's pleasure and glee are not dependent upon birth in the high status family, beauty, speech, intellect, appearance and devotion to Him. Bhagwan Shri Ram loved forest dwellers who were devoid of these qualities.*

Similarly, in the 2nd chapter of the 11th *Skandh*, the Bhagwan says to King Nimi :

न यस्य जन्म कर्मभ्यां न वर्णाश्रमजातिभिः।
सज्जतेऽस्मिन्नहंभावो देहे वै स हरेः प्रियः।। 11,2,51

*He is only dear to Hari who never embellishes his body with the ego of his birth, Karm, Varn, Ashram or Caste. It simply means that a person who takes pride in his clan, Varn, etc., can never be dear to Bhagwan.*

न यस्य स्वः पर इति वित्तेष्वात्मनि वा भिदा।
सर्वभूतसमः शान्तः स वै भागवतोत्तमः।। 11,2,52

*A great devotee of Bhagwan is one who does not differentiate between himself and others in matters of wealth and body and who considers all beings as equal and remains at peace.*

He, who accumulates resources or wealth in excess is a thief by instinct. The Bhagwan says :

यावद्भ्रियेत जठरं तावत् स्वत्वं हि देहिनाम्।
अधिकं यो अभिमन्येत स स्तेनो दण्डमर्हति।। 7,14,8

*Only that much food with which the stomach can be filled is right for living beings. Anyone who wants more than that is a thief and is entitled to punishment. (7,14,8)*

Could there be a more revolutionary idea of egalitarianism than this? In the 6th chapter of the *Bhagwat Puran* itself, the *Purankar* has expressed the social equality. The concern for not only human beings but also for all beings is visible in this chapter.

Yudhishthira's Rajasuy *Yagy* is described in the 74th chapter of the 10th *Skandh* of the *Bhagwat Puran*. The following verse clearly describes that the four *Varns* of the society were respected equally :

ब्राह्मणः क्षत्रियः वैश्याः शूद्राः यज्ञदिहदृवः।
तत्रेयुः सर्वराजानो राज्ञां प्रकृतयो नृप।।

*Many eminent Rishis, Kings, Brahmans, Kshtriys, Vaishys, and Shudrs were present to witness the Rajasuy Yagy.*

The above *Shlok* shows that *Yagy* etc. were public and religious events in ancient times. All the *Varns* took equal part in the event and there was no disrespectful discrimination against anyone.

The *Purankar* told the devotee in the 7th chapter of the 7th *Skandh* :

हरि सर्वेषु भूतेषु भगवानास्त ईश्वरः।
इति भूतानि मनसा कामैस्तैः साधु मानयेत्।।

*The Bhagwan resides in all beings. With this feeling and respect towards all beings have respect for all, try to fulfill their wishes as much as possible.*

The above utterance of the *Bhagwat Puran* is a manifestation of the declaration of the statement of the *Vedic* literature 'आत्मवत्सर्वभूतेश' (*everyone is like my Aatma*). This comprehensive and all-touching statement of the *Bhagwat Puran* is the ideal conception of an egalitarian society, in which there is no trace of discrimination against creatures. *Muni* Maitreya declares the importance of purity of character in the 30th chapter of the 4th *Skandh* of the *Bhagwat Puran* :

य इदं सुमहत् पुण्यं श्रद्धयावहितः पठेत्।
श्रावयेच्छृणुयाद् वापि स पृथोः पदवीमियात्।।
ब्राह्मणो ब्रह्मवर्चस्वी राजन्यो जगतीपतिः।
वैश्यः पठन् विट्पतिः स्याच्छूद्रः सत्तमतामियात्।।

*One who reads, hears or narrates this most sacred Prithu Charit with reverence attains the position of King Prithu, i.e. the Supreme abode, Brahmans attain Brahman glory, Kshtriys attain the kingdom, Vaishys become the best amongst merchants and Shudrs become superior.*

It is clear from this example that all had equal rights to read and recite the *Bhagwat*. A *Shudr* was also entitled to its discourse. Clarifying it further, it is declared in the *Bhagwat Puran* that :

विप्रोऽधीव्याप्नुयात् प्रज्ञां राजन्योदधिमेखलाम्।
वैश्यो निधिपतित्वं च शूद्रः शुद्धयेत पातकात्।। 12,12–64

*By reading this Bhagwat, Brahman attains knowledge, Kshtriy gets the earth surrounded by sea, Vaishy becomes the owner of wealth and all the funds, the Shudr is freed from all sins and becomes pure. (12,12-64)*

Even after going through these revolutionary pronouncements, some scholars appear to be stubbornly saying that *Shudrs* do not have the right to read the *Purans*. They only have the right to listen to its discourse. In fact, these scholars are bent upon and are engaged in subverting *Bhartiy* culture. Though on this subject there are several clear proclamations in the *Bhagwat* yet these dishonest and mischievous intellectuals have put on spectacles of misconceived ideas. They do not want to understand these sublime proclamations of the *Bhartiy* scriptures. That is why by imposing their own contrasting meaning they make foolish interpretations. To dispel their darkness and ignorance the following verse of *Bhagwat* is enough :

एतद् वै सर्ववर्णानामाश्रमाणां च सम्मतम्।
श्रेय सामुत्तमं मन्ये स्त्रीशूद्राणां च मानद।। 11,27,4

*In this context Bhagwan Shri Krishn says to Uddhava - This ritual is valid for all Varns and all Ashrams. I consider these equally beneficial for women and Shudrs.*

Astu! I feel above quotes of the *Bhagwat* are splendid pillars of universal equality. By eliminating the darkness of distortions and misinformation prevalent in the present society in the light of these splendid pillars, all acceptable paths can be paved.

# Bhavishy Puran

The *Bhavishy Puran* is an authentic and revolutionary text to understand the *Vedic* form of the *Varn System*. The *Bhavishy Puran* completely and firmly refutes birth-based caste system. According to a few people as there is a difference between a horse and a donkey, there is a difference between a *Brahman* and a *Shudr*. According to the *Bhavishy Puran*, there are no physiological differences like composition, shape, colour, blood and bone etc. between a *Brahman* and a *Shudr*. According to the *Purans*, *Shudrs* have also been given rights to study the *Veds* and *Vedant* and to perform *Yagy*. All human beings are inherently equal. In the 44th chapter of the *Brahmaparva* of the *Puran*, it is clearly propounded that the distinction between the species and sub-species of animals such as elephants, horses and donkeys is not found between *Brahmans*, *Kshtriys*, *Vaishys*, *Shudrs*. Therefore the birth is not a justification for *Varn* determination. There is no difference in the shape and color even among people of different *Varns*. According to the *Purankar*, *Brahmans* who do inferior business, eat non-edible foods, adopt objectionable attitudes and are characterless cannot be considered *Brahmans*. Hence the *Varn* cannot be determined on the basis of birth :

गोवर्गमध्यं च गतो यथाश्वो निधर्यिते ज्ञैः सुविचक्षणत्वात्।।
मनुष्यभावादविशिष्यमाणस्तद्वद् द्विजः शूद्रगणान्न भिन्नः।। 40/20

*Any knowledgeable person can easily distinguish and identify a horse amongst cows, ox and bulls. However, there is no such distinction between groups of humans, Brahmans or Shudrs. These cannot be identified or differentiated amongst the group of human beings. (40/20)*

The identity of any *Varn* by birth ends the moment Gun and *Karm* get associated with that *Varn*. The *Puran* propounds this principle in the following way :

किं ब्राह्मणां ये सृकृतं त्यजन्ति किं क्षत्रिया लोकमपालयन्तः।

स्वधर्महीना हि तथैव वैश्याः शूद्राः स्वमुख्यक्रियया विहीनाः।।

भवि० पु० 40,33

*What Brahman is he who has given up virtue; what Kshtriy is he who does not take care and nurture his subjects; what Vaishy is he who is devoid of his Dharm and what Shudr is he who has become devoid of his main activities. (Bhavishy Puran 40.33)*

In the next verse the *Purankar* describes the caste system as forbidden and imaginary by declaring it as propounded by fools :

एवं प्रमाणैः प्रतिसिध्यमानां साङ्केतिकी याति नरो व्यवस्थाम्।
स्वकीयसिद्धां स्वमतैर्निषिद्धां न बुध्यते मूढमना वराकः।।

*That is, in this way the caste system of Brahmans, Kshtriys, Vaishys and Shudrs among human beings is forbidden by the evidence and it is only an imaginary system. Foolish people do not understand that this system is made by themselves and is prohibited according to their own opinion to be forbidden.*

There are many ways in which a person can give up his *Varn*. The *Puran* says :

शूद्रो ब्राह्मणतामेति ब्राह्मश्वैति शूद्रताम्।
क्षत्रियो याति विप्रत्वं विद्याद्वैश्यं तथैव च।। भवि० पु०

*Shudr can become a Brahman and a Brahman can become a Shudr. And that Brahman, Vaishy, Kshtriy or Shudr who do not follow the attributes that come under that Varn whose attributes they start following. (Bhavishy Puran)*

Manu believes that *Shudrs* can attain *Brahmantav*; *Brahmans* can also become *Shudrs*; *Kshtriys* and *Vaishys* can also attain *Brahmantav* or they can become *Shudrs*.

In the 41st chapter of the *Bhavishy Puran*, the *Shudrs* have been given the right to recite *Vedic Mantrs*, perform *Sandhya-Vandan* and other rituals; they can bind their crest (Shikha); pray, tie mekhla (the girdle near navel), carry a *Dand* (symbol of renunciation); can wear deer skin etc. The *Purans* say that even when the highest Gods are not able to deprive *Shudrs* of their rights, how can human beings do that? The *Purankar* has made it clear that when there is no difference between the senses, appearances, basic instincts, etc. of *Brahmans* and *Shudrs*, how can *Shudrs* be different from *Brahmans*? Just as there is no caste distinctions amongst four sons of a father, there cannot be any difference between the sons of the Supreme Father or the Supreme *Aatma*. If all the verses mentioned in this chapter had been properly perused and understood and accordingly put before the society, surely there would not have been any disintegration in society. Some verses are worthy to be paid attention :

शिखाप्रणवसंस्कार सन्ध्योपासनमेखलाः ।
दण्डाजिनपवित्राद्याः शूद्रेष्वपि निरङ्कशाः ।। भवि० पु० 41,10

*Shudrs are absolutely free to bind their crest, recite Vedic Omkar Mantr, perform sixteen sanskar (sixteen rites), do Sandhya-Vandan, Upasana, wear Mekhla near navel and wear deer skin, carry Dand (symbol of renunciation) and can use other purifying materials.*

प्रसङ्गनेऽपि हि शूद्राणां न शक्यो विनिवारितुम।
देवोत्तमंत्रयेणापि निवर्तन्ते नरा स्वयम् ।। भवि० पु० 41,11

*Even the trinity of Brahma, Vishnu and Mahesh are not able to stop the involvement of Shudrs in the above referred activities. If the Hindu trinity cannot interfere in these rights of the Shudrs, no human being has the capacity to do so.*

चत्वार एकस्य पितुः सुताश्व तेषां सुतानां खलु जातिरेका।
एवं प्रजानां हि पितैक एवं पित्रैक भावान्न च जाति भेदः ।।

भवि० पु० 41,45

*Just as a father has four sons, but the same caste is ensured for those four sons, how can there be any caste differences amongst all the children of the same father may be, i.e. Purush, Narayan or Brahm??*

This chapter abounds in similar revolutionary and vigorous arguments.

The 42nd chapter of the *Bhavishy Puran* also expresses similar revolutionary ideas. In this chapter, those who claim to be from superior *Varn* are ridiculed fiercely and called intoxicated and eunuch, as there is no difference between a *Brahman* devoid of *sanskars* or any other person. The eyes of those who see caste discrimination are capricious. In this chapter there is a reference to *Rishis* like Vyas, Parashar, Shukdev, Kanad Shringrishi, Vashishth, Madanpal and Mandavy, who were not from *Brahman* clan but became superior *Brahmans* because of their knowledge and wisdom.

In the 43rd chapter, refuting caste discrimination, it has been said that even those who perform all rituals may be found to be immoral. Due to the absence of specific attributes in some persons in the four *Varns*, there has always been crossings over one *Varn* to another. Therefore, the claim of *Varn* superiority is mere arrogance.

In the 44th chapter, instead of *Varn* gradation, the principle of *Varn System* was propounded based on distinctive attributes and performance of duties related to that particular *Varn*. Just as the *Shrimad Bhagwad Gita* describes the natural actions of the four *Varns*, similar thoughts are found in this chapter of the *Bhavishy Puran* too. It is clearly declared in the chapter that any sinless person is a *Brahman*, and a virtuous *Shudr* is above a *Brahman* and a *Brahman* without character is most inferior.

According to the *Purans*, *Brahmantav* grows rapidly in a person who has virtues like forgiveness, control over senses, compassion, charity, chastity, purity, honesty, patience, graciousness, softness, simplicity, contentment, absence of ego, austerity, modesty, courage, wisdom, celibacy, scholarship, meditation, non-maliciousness, detachment, peace, freedom from jealously and craving and exhibits, self-control over speech and body. Such splendid and great persons who have become pure in mind because of the distinctive attributes and conduct acquire the respective *Varn* due to their pious thoughts. (*Bhavishy Puran* 44.12.15)

In relation to the above types of persons the *Puran* declares that :

मन्वमन्तरेणु सर्वेषु चतुर्युगविभागशः ।
वर्णाश्रमाचारकृतं कर्म सिद्धयत्यानुत्तमम् ।। भवि० पु० 44,17

*Such scholars who accept these thoughts are the true scholars of the Puran, Veds, and have profound knowledge of the Gita and the Smritis are gentlemen. In all the Manvantaras, according to the division of the four Yugas, the best accomplishment is achieved through actions performed according to the attributes ascribed to Varnashram.*

मै वै परिगृहीतारस्तेषां सन्त्वबलाधिकाः ।
इतरेषां क्षतत्राणान् स्थापयामास क्षत्रियान् ।।

उपतिष्ठन्ति ये तान्वै याचन्तो नर्मदा सदा।
सत्य ब्रह्म सदाभूतम् वदन्तो ब्राह्मणास्तुते।।
ये चान्येऽप्यबलास्तेषां वैश्य कर्मणि संस्थिताः।
कीलानि नाशयन्ति स्म पृथिव्यां प्रागतन्द्रिताः।।
वैश्यानेव तु तानाह कीनाशान्वृत्तिमाश्रितान्।
शोचन्तश्च द्रवन्तश्च परिचर्यासु ये नराः।।
निस्तेजोऽल्पवीर्याश्च शूद्रांस्तानब्रवीन्तु सः।।

भवि० पु० 44,20–23

*This means that those who, being more powerful, will perform the responsibility of saving others from the destruction will be called Kshtriys. Those who in all humbleness preach truth and eternity of Brahm to these Kshtriys will be called Brahmans. Those amongst them who are not strong enough will continue to do the deeds according to Vaishy Varn. In the ancient times, the one who gave up laziness, uprooted trees on the earth and made it cultivable and also those who do the works of such as farmer will be called Vaishy. Despite being in a doleful and in poor condition, those who serve the above three Varns and are weak and have little power, will be called Shudrs.*

ब्राह्मणक्षत्रियविशां शूद्राणां च परस्परम्।
कर्माणि प्रवि भक्तानि स्वभाव प्रभवैर्गुणैः।। भवि० पु०

*Thus, the actions of Brahmans, Kshtriys, Vaishys and Shudrs become inherently different just by their natural qualities.*

In the verses from 25 to 32 of the same chapter, the behavioural deeds of *Brahmans*, *Kshtriys*, *Vaishys*, *Shudrs* have been explained, on the basis of which the *Purans* clearly state in the 33rd verse that there is only one caste of human beings and the different caste patterns which came into existence at a later stage were based on birth and not on distinctive attributes and *Karm*. Caste is a practical expression, but the nature of each *Varn* is different due to its distinctive attributes and tendencies of a person. The human race, which is pervasive in the totality of feeling, difference, death, birth and condition, is not comprehensible to those who are always ready to build the dividing wall of caste among human beings.

यद्येका स्ुटमेव जातिरपरा कृत्यात्परं भेदिनी,
यद्वा व्याहृतिरेकतामधिगता यच्चान्यधर्मययो।

एकैकाखिलभावभेदनिधनोत्पत्तिस्थितिव्यापिनी,
किं नासौ प्रतिपत्तगोचरपथं यामाद्विभक्त्या नृणाम्।।

भवि० पु० 44,43

*Thus, the five chapters of the Bhavishy Puran from the chapters 40 to 44 describe the great edifice of the Varn System. These five chapters are a slap on the cheeks of those who want to tarnish Varn System to be based on birth.*

After going through many verses from the *Bhavishy Puran*, a lofty and noble picture of the *Bhartiy Varn System* is clearly visible. There is an urgent need to bring this truth to the forefront. This is our *Yugdharm*.

# References from Upanishads

According to *Upnishads* the *Varn System* is determined not by birth but by virtue and action. A few of the evidences are given in the following pages :

Radhakrishnan, the great thinker and teacher of Bharat, has given an important place to the *Vajrsuchay Upanishad* among the eighteen major *Upnishads. The Vajrsuchay Upanishad* comes under *Samved.* The ignorant treat it as pollution and the wise consider it as a jewel. This revolutionary book of Indian social system was deliberately not given importance, but since 'the truth always asserts', its revolutionary ideas are now becoming known to everyone.

According to the *Vajrsuchay Upanishad* :

1. ब्राह्मणक्षत्रियवैश्यशूद्रा इति चत्वारो वर्णास्तेषां ब्राह्मण एव प्रधान इति
वेद वचनानुरूपं स्मृतिभिरप्युक्तम्।
तत्र चोद्यमस्ति को वा ब्राह्माणो नाम?
किं जीवः, किं देहः, किं जातिः, किं वर्ण, किं पाण्डित्यं,
किं धर्मः, किं ज्ञानं, किं कर्म, किं धार्मिक इति।।

*There are four Varns : Brahmans, Kshtriys, Vaishys and Shudrs. In them, according to the words of Veds and Smriti Shastrs, Brahmans are the main ones. The question to be considered here is who is a Brahman? Is Aatma Brahman by body, caste, Varn, knowledge, Karm, Dharm or righteousness.*

तत्र प्रथमो जीवो ब्राह्मण इति चेत् तन्न।
अतीतानामत वर्तमानानेकदेहानां जीवस्यैकरुपत्वाद्।।
एकंस्यापिकर्मवशाद्नेक देहसम्भवात्।
सर्वशरीराणां जीवस्यैक रूपत्वाच्च।
तस्मान्न जीवो ब्राह्मण इति।।

*The first living being i.e. the Aatma cannot be a Brahman, because in the past, present and future one Aatma resides in many bodies. That is to say, the Aatma, being immortal, assumes the new forms as a living being after death of previous embodied forms. Although the Aatma is one, yet due to the cycle of Karm, it takes rebirth in new bodies. These may be of a Brahman, a Chandal, an animal or a bird. One Aatma may reside in several living entities over periods of time. If someone's Aatma was a Brahman, he would be a Brahman in every birth. But in reality it does not happen like that. So the Aatma is not a Brahman.*

2. नर्हि देहो ब्राह्मण इति चेत् तन्न।
आचाण्डलादिपर्यन्तानां मनुष्याणां पाञ्चभौतिकतत्वेन देहस्यैक
रूपत्वांञ्जारामरण धर्मा धर्मादि साम्य दर्शनात्।।
ब्राह्मण श्वेत वर्णः, क्षत्रियोंः रक्त वर्ण,
वैश्यः पीत वर्णः शूद्रः कृष्ण वर्णः
इति नियमाभावात्।
पित्रादिशरीरदहनें पुत्रादीनां ब्रह्म हत्यादि दोषसम्भवाच्च।
तस्मान् देहो ब्राह्मण इति।।

*Then is the body a Brahman? Never, because from the Chandal to Brahman all human beings are homogeneous. What is the reason for this homogeneity or similarity? Brahmans, Kshtriys, Vaishys and Shudrs all are made of the five elements like Earth, Water, Fire, Space and Vayu. It is not that Brahmans have more elements or Chandals have lesser. Everyone from Brahmans to Chandals have the same body : one head, two eyes, two ears, one nose, two hands, two legs etc. It is not that the Brahmans who are born from the mouth of Brahm have four mouths and the Shudrs who are born from the feet of Brahm have no mouth at all. Birth and death which are eternal are also the same from Brahmans, Kshtriys, Vaishys, Shudrs. All die. Therefore, there is no determination of Varn on the basis of body. If this body was a Brahman, the cremation of parents by son or daughter would be taken as a sin equivalent to that of killing of a Brahman. In practice it does not happen. In some religious texts it is said that Brahmans are white, Kshtriys are red, Vaishys are of yellow and Shudrs are black in colour. Such a rule does not appear in reality. Brahmans may be black and Shudrs may be white by complexion. Therefore, by just looking at the body, Varn cannot be determined.*

3. तर्हि जातिर्ब्राह्मण इति चेत तत्र

तत्र जात्यन्तजन्तुष्वनेक जाति सम्भवा महर्षयो बहवः सन्ति।।
ऋष्यशृङ्ग मृग्याः कौशिकः कुशात्,
जाम्बूको जम्बूकात, बाल्मीको बाल्मीकात्।
व्यासकैवर्तकन्यकायाम, शशपृष्ठाद गौतमः वशिष्ठ उर्वश्याम।।
नारदो दासी पुत्रः, अगस्त्यः कलशे जात इति श्रुतत्वात्।
एतेशां जात्या विनाप्यग्रे ज्ञानप्रतिपादिता ऋषयो बहवः सन्ति।।
विनापि सम्यग ज्ञानविशेषाद् ब्राह्मण्यमत्युत्तमम् श्रूयते।
तस्मान्न जाति ब्राह्मण इति।।

*Is there any one Brahman by caste (birth)? It is not so. Many beings outside the human race and many persons born in castes other than the Brahman caste have become great Maharishis. It is heard that Rishi Shring was born from deer, Kaushik from Kush, Jambuk from jackal, Valmiki from termite mound, Vyas from sailor's daughter, Gautam from rabbit's back,* Vashishtha *from prostitute Urvashi, Narad from domestic maid and Agastya from an urn. These Maharishis attained the position of the best Brahmans without being born in Brahman caste. They acquired the Brahman Varn just by the propounding proper knowledge and wisdom. On the basis of this, these Maharishis not born in a Brahman clan were still considered to be the best of the Brahman Maharishis. This proves that no one becomes or is a Brahman by caste.*

Similarly, the *Upnishads* say that there is no *Brahman* by knowledge, deeds, erudition or religious deeds. The question is, who is a *Brahman*? The *Upnishads* also give the answer to the question.

4. य कश्चिदात्मानमद्वितीयं जातिगुण क्रियाहीनं षड्र्मिषड्भावेत्यादि
सर्वदोषरहितं सत्यज्ञानानन्दानन्तरूपरूपं स्वयं
निर्विकल्पम् अशेषकल्पाधारम् अशेषभूतान्तर्यायित्वेन
वर्तमानमन्तर्बहिश्चाकाशवदनुस्यूतम् अखण्डानन्दस्वभाववम् अप्रेमयम्
अनुभवैकवेद्यम् अपरोक्षया भासमानं करतलामलकवत् साक्षात्
अपरोक्षीकृत्य कृतार्थतया कामरागादिदोषरहितः शमादिगुणसम्पन्नो
भावन्मात्सर्यतृष्णाशामोहादिरहितो दम्भाहङ्कारादिभिः
असंस्पृष्टचेता वर्तत् एवमुक्तलक्षणो यः स एव ब्राह्मण
इति श्रुतिस्मृति पुराणेतिहासानामभिप्रायः अन्यथा हि
ब्राह्मणत्व सिद्धिर्नास्त्येव

*A Brahman is a person who is not affected by six senses like six waves; who is cast in a distinctive mould of attributes and actions; who is the eternal form of truth, knowledge, bliss and is without any choices, Puran Kalpadhar, present in all beings, is of unbroken bliss like pervading sky and who from his immeasurable experiences knows the knowable and oberserable Aatma and is devoid of sensual flaws and defects like saam- daam, jealousy, cravings, hope, infatuations, arrogance and ego. A person infused with such qualities and lacks defects is a Brahman. Otherwise Brahmantav cannot be there. This is the essence of the Veds, Dharamshastr, Purans and historical texts.*

Finally the *Upanishads* proclaimed that :

किन्तु करतलामलकमिव यः पश्यत्यपरोक्षेण कृतार्थः।
काम रागद्वेषादि रहितः शमदमादिसन्तोषो मानमात्सर्य तृष्णा सम्मोहादि
दुष्टार्थनिवृत्तः स एव ब्राह्मण उच्चते।।
जन्मना जायते शूद्रो व्रत बन्धाद् द्विज उच्चते।
वेदाभ्यासी भवेद् व्रिपो ब्रह्म जानाति ब्राह्मणः।।
अत एव ब्रह्मविदेव ब्राह्मणो नान्य इति निश्चयः

*He is called a Brahman who perceives the divine like a gooseberry in his hand and is free from defects like lust, passion etc.; who is satisfied with the qualities of saam-daam etc. and has got rid of the vices like pride, jealousy, craving, attachment, etc. All are Shudrs by birth. They owing to their distinctive attributes and Karm, are reborn in another Varns. Therefore, they are called Dwijs. By practicing the Veds, one is considered Vipr and one who knows Brahm is a Brahman.*

# References from Smritis

Today, the *Manusmriti* is the most discussed of the *Smritis*. It is also a fallacy that the *Manusmriti* was written by Bhagwan Manu, the creator of the universe after the *Pralay*. Before discussing the references to the *Manusmriti*, it needs to be understood that the *Manusmriti* was not created by Bhagwan Manu, who is considered as the *Aadi Purush*. In fact, it was composed much later. Maharaja Manu had raised the prestige of India on the whole earth by giving excellent governance and praise worthy social justice but due to some narrow principles interpolated later in the *Manusmriti*, *Manuwad* is being used as an abuse. If one does a proper study of the character of Maharaja Manu, it will be evident that King Manu was so generous that he allowed two sons to become *Brahmans*, one to become *Vaishy* and one to become *Shudr*, and allowed one of his sons to become a forest dweller, and deprived one of his sons of the state property due to his malfeasance.

Dr. Ambedkar has also proved that the author of the present *Manusmriti* is not *Adi Purush* Manu but a *Rishi* named Sumati Bhargav. According to Dr. Ambedkar, in the history of ancient Bharat, the name of 'Manu' was a sign of respect, so for the purpose of giving glory to this code, Manu has been described as its writer because such a practice was prevalent in the ancient times. This *Samhita* is associated with Bhriguvansh. The actual name of this *Samhita* of Bhrigu is 'Manavdharm Samhita'. The name of Bhrigu is added at the end of each chapter of this *Samhita*. In this book itself, we get information about the clan of the author of the *Samhita* but the personal name of the author is not mentioned in it. Many writers were aware of this. Around fourth century, the author of the *Narad Smriti* knew the name of the author of the *Manusmriti*. According to the Narad, a person named Sumati Bhargav composed the *Manusmriti*. Sumati Bhargav is not a fictitious name; he must have been a historical person. This is because the great commentator on the *Manusmriti*, Medhatithi, was of the view that this Manu was actually a person. Thus, Manu was the surname or pseudonym of Sumati Bhargav and this Sumati Bhargav was its real author.

The above discussion of Dr. Ambedkar is also supported by the *Narad Smriti*. Not only this, it is also mentioned in the so-called *Manusmriti* that once many *Maharishis* paid obeisance and worshiped Bhagwan Manu, who was meditating and requested him 'O Bhagwan!' It is a humble request to you to kindly tell us the *Dharm* to be practiced by the people of the four *Varns Brahman*, *Kshtriy*, *Vaishy*, *Shudr* and *Varnasankar* (children born by the combination of male of *one Varn* and female of different *Varn*). In this context this verse of the *Manusmriti* is worthy of attention:

मनुमे काग्रमासीनमभिगम्य महर्षयः
प्रतिपूज्य यथा न्यायमिदं वचनमब्रुवन।
भगवन्सर्ववर्णानां यथावदनुपूर्वशः।
अन्तरप्रभावाणां च धर्मान्तो वक्तुमर्हसि।।

*Thereafter, Bhagwan Manu started describing the creation to the Rishis. According to the Manusmriti, Bhagwan Manu himself spoke to Rishis till 62[nd] verse of the first chapter of the Smritis. In the 62[nd] verse of the first chapter of Smriti itself, Bhagwan Manu said that 'Maharishi Bhrigu, one of the Rishis like Marichi, received the knowledge of this scripture from me. Maharishi Bhrigu will narrate this scripture to other Rishis.*

एतद्वोऽयं भृगुः शास्त्रं श्रावयिष्यत्यशेषतः।
एतद्धि मत्तोऽधिजगे सर्वमषोऽखिलं मुनिः।। मनुस्मृति 1,62

*The disputed present Manusmriti also clarifies that the statement in the Manusmriti or the announcement made by Maharishi Bhrigu in front of other Rishis. In the Narad Smriti, this Bhrigu is mentioned as Sumati Bhargav. The Narad Smriti mentions that:*

इह हि भगवान मनुः प्रथमं
सर्वभूतानुग्रहार्थमाचारस्थितिहेतुभूतं शास्त्रं चकार।
सुमतिरपि भार्गवस्तस्मादर्धात्य तथैवायुर्हासादल्पीयसी
शक्तिर्मनुष्याणामिति चतुर्भि सहस्त्रैः संचिक्षेप

तदेतत्पितृमनुष्या ह्यधीयन्ते विस्तरेण
शतसहस्त्रं देव गन्धर्वादयः।।

*That is, first of all, Bhagwan Manu composed a scripture of one lakh verses for the benefit of the living beings and gave this scripture to Narad. Narad, after finding this voluminous scripture, condensed it into 12000 verses and handed it over to Maharishi Markandeya and he also reduced it further and converted it into 8000 verses and handed it over to Sumati Bhargava. Sumati Bhargava studied it and reduced it to 4000 verses and since then its study was started by our ancestors and humans.*

This quote from the *Narad Smriti* at least makes it clear that the book which is called the *Manusmriti* was given to everyone by a *Rishi* named Sumati Bhargav and not by Manu Maharaj himself.

In this context, it is also necessary to understand that the successor, who occupies the chair of any Jagadguru Shankracharay is called Jagadguru Shankracharay. In the same way many *Peeths* were established in the past. Whoever sat on the seat of the *Peeths*/Chair would get the same title. Similarly, Parshuram appears in *Tretayug* and also appears as Karan's guru in *Dwaparyug*. Similarly, Hanuman appears with Ram in *Tretayug*, while in the *Mahabharat* period, the same Hanuman forces Bhim to remove his tail. After studying the *Bhartiy* theology, there is no illusion that this person is not an individual but is the successor who took care of the chair established by his predecessors but was addressed by the name of the original person. It is also clear from the study of the scriptures that in the same way a Chair was established in the name of Swayambhu Manu, who was considered as the *Adi Purush*. And after this, a series of successive Manus appeared. In the beginning of the present *Manusmriti*, there is a discussion of six Manus, descendants of Manu. There is one more thing 'Manu' has been held in great respect and reverence in *Sanatan Dharm*. By the later writers, thinking that if anything is put in front of the society in the form of Manu's statement, the society will accept it easily. That is perhaps why several statements appeared in the name of Manu in many later texts. The so-called *Manusmriti* may also have been founded on this very thought. The great scholar Shri P.V. Kale has written on page 310 of his book the *History of Dharamshastr* that it is almost impossible to say who is the author of the *Manusmriti* but It is so indisputable that Manu, the father of human beings, cannot be its creator.' According to Kale, 'What would have been the purpose behind declaring Manu as its author by the unknown author is difficult to say, but glorifying the book must have been an objective.' Shri P.V. Kale has said on page number 326 of this book that considering *Manusmritikar* to be the son of *Brahma* and the father of human community is a camouflage because the present *Manusmriti* mentions later human writers like Atri, Utthya-Tanay, Shaunak, Bhrigu and Vashishth.

The names of such *Rishis* who were born much later are mentioned in different chapters in the *Manusmriti*. If we examine the *Manusmriti* properly, then we find many such verses in which discussion of castes till modern times is visible, which makes it clear that the *Manusmriti* was not composed by Adi *Purush*, but by someone else much later on. Many verses were added later in the *Manusmriti*, which destroyed the original content of the *Manusmriti*. A verse from the 10th chapter of the *Manusmriti* says:

पौण्ड्रकाश्चौड्रद्रविडाः काम्बोजा यवनाः शकाः।
पारदापहलवाश्चीनाः किराताः दरदाः खशाः।।

The mention of Paundrak, Aundra, Dravid, Kamboj, Yavana, Shaka, Parad, Palhava, Cheen, Kirat, Darad and Khasha in this verse is sufficient to make it clear that the *Manusmriti* was not composed by Adi *Purush*. At the same time, it also seems that due to selfishness and conspiracy, many verses were either added or removed at different joints of time. If the so-called persons of social influence and erstwhile priests did not find any verse suited to their feelings or interest, some of these were deleted and for wish fulfillment or exploitation, some verses were composed and added to the original *Smriti*. While interpolating new verses, these priests forgot to omit the previous verses. That is why the *Manusmriti* is full of contradictory verses. Whether it is the issue of meat-eating or inter-caste marriage or *Niyog* or *Dayabhag*, contradictory verses can be found all over. A German scholar, George Bühler said that, "In the beginning there was a human theology, which was changed over time, modified and given the form of the new present *Manusmriti*." In fact, the present form of the *Manusmriti* itself is becoming a subject of criticism. Most of the text of the *Manusmriti* is exemplary and ideal. But there are many such verses as will not be acceptable to any civilized society.

Therefore, it is incumbent upon the social reformers and organizations to put a joint effort so that interpolated portions and utterances added with malafide intentions in the scriptures and the other texts should be removed from these *Dharmik* scriptures, and by unanimously accepting the ideal principles of social formation, the modified ideal versions should be presented in front of the general public. This would integrate and harmonize the *Bhartiy* society and pave the best path for the world.

If we observe the *Manusmriti* in the above light, then we will find many *Shloks* which the society can proudly proclaim as the pillars of the *Bhartiy* culture. Some *Shloks* are given below as examples :

सत्यं ब्रूयात् प्रियं ब्रुयान्नब्रुयात् सत्यमप्रियम्।
प्रियं च नानृतं ब्रूयादेश धर्मः सनातनः।। 4,138

*Speak the truth, speak the pleasant, do not speak unpleasant truth and pleasant untruth. This is Sanatana Dharm. (4,138)*

धृतिः क्षमा दमोऽस्तेयं शौचमिन्द्रियनिग्रहः।
धीर्विधा सत्यऽक्रोधो दशकं धर्मलक्षणम्।। 6,91

*Patience, forgiveness, strength, non-stealing, defecation, control over senses, knowledge, learning, truth and non-anger are the ten characteristics of Dharm. (6,91)*

धर्म एव हतो हन्ति धर्मो रक्षति रक्षितः।
तस्माद्धर्मो न हन्तव्यो मा नो धर्मो हतोऽवधीत्।। 8,15

*Whoever destroys Dharm, Dharm destroys him completely. Dharm protects the one who protects it. So never destroy Dharm so that a protected Dharm can protect us. (8,15)*

सभां वा न प्रवेष्टव्यं वक्तव्यं वा समञजसम्।
अब्रुवन विब्रुवन् वापि नरो भवति किल्वषी।। 8,13

*Either do not go to a gathering or if you go, then speak only what is right. A liar and a silent person is a commissary of sin. (8,13)*

न जातु कामः कामानामुपभोगेन शाम्यति।
हविषा कृष्णवर्त्मेव भूय एवाभिवर्धते।। 2.97

*Kaam is never pacified by the enjoyment of sensual objects, but is enflamed more and more by these sensual objects. These make the flames to rise higher and higher. (2,97)*

श्रद्धानः शुभां विद्यामाददीतावरादपि।
अन्त्यादपि परं धर्मं स्त्रीरत्नं दुष्कुलादपि।। 2,242

*Knowledge should be gained with reverence even from a person inferior to oneself. Excellent Dharm should be obtained from the Chandal and also a girl with auspicious signs should be accepted for marriage even if she belongs to a low family. (2.242)*

Therefore, it is clear from the impartial study of the *Manusmriti* that the *Smritikar* has expressed very broad and sublime views on subjects like *Dharm*, ethics etc., which are worthy of adopting with a sense of pride by any civilized society. I have said earlier that the *Manusmriti* contains many verses which were added from time to time later and many verses were omitted in many versions. Both these actions appear to be the result of a well thought-out morbid scheme. *Smritis* are texts on behaviour prescribed for a particular time. It is clearly mentioned in these *Smritis* that if the word or statement of the *Smriti* is contradictory with that of Shruti i.e. the *Veds*, then the statement of *Smriti* will automatically be considered null and void. According to the verse quoted in the fourth chapter of the *Manusmriti*.

परित्यजेदर्थकामौ यौ स्यातां धर्मवजितौ।
धर्मं चाप्यसुखोदर्कं लोकसङ्क्रष्टमेव च।। 4,176

*If Arth and Kaam are taboo according to Dharm then these should be abandoned. Similarly any Dharmik law or practice in the course of time becomes painful and considered insignificant by people, it should be abandoned.*

A similar view is expressed in the *Yagynavalkya Smriti* :

कर्मणा मनसा वाचा यत्नाद् धर्मं समाचरेत्।
अस्वर्ग्यं लोकविद्विष्टं धर्मामप्याचरेन्न तु।।

*One should not only practice Dharm diligently by mind, word and deed, but also do not practice such Dharm which does not lead to heaven and is detested by the people.*

It is clear from the above quotations of both the *Smritis* that anything against the public interest should not be acceptable. Surely it should be abandoned.

The *Yajnvalky Smriti* gives the right to the *Shudrs* to perform the five *Yagys*

भार्यारतिः शुचिर्भृत्यभर्ता श्राद्धक्रिया परः।
नमस्कारेण मन्त्रेण पञ्चयाज्ञान्नहापयेत्।। 1.121

*A Shudr engaged in sex with wife, purity, upkeep of servants and is performing Shradh should never give up PanchaYagy with the prescribed Namaskar Mantr.*

According to *Yajnvalky Smriti*, these five *Yagyas* are as follows:

बलिकर्मस्वधाहोमस्वाध्यायातिथि सत्क्रियाः।
भूतपित्र्यमरब्रह्ममनुष्यां महामखाः।

*Balikaram, Swadha, Hawan, Self-study are said to be great Yagy for mankind. These five Yagys are called as Bhoot Yagy, Pitri Yagy, Dev Yagy, Brahm Yagy, etc.,* as-

धमेप्सवस्तु धर्मज्ञाः सतां वृत्तिमनुष्ठिताः।
मन्त्रवर्ज्याः न दुष्यन्ति प्रशंसां प्राप्नुवन्ति च।।

*Those Shudrs who are interested in the works of Dharm and know the principle of Dharm, perform Yagy even without the knowledge of Mantr are not condemned but praised.*

According to the *Manusmriti*, at the time of creation of the universe, *Brahma* first created the *Viraat Purush*, the *Viraat Purush* created Bhagwan Manu by performing *Dharmik* austerities. Bhagwan Manu produced ten *Maharishis* by performing *Dharmik* austerities. These *Maharishis* are - Maricha, Atri, Angira, Pulasty, Pulah, Krutu, Prachet, *Vashishth*, Bhrigu and Narad.This is clear from verses 36 and 37 of the first chapter of the *Manusmriti*. The lineage of *Brahmans*, *Kshtriys*, *Vaishys* and *Shudrs* on the basis of the *Manusmriti* is as follows:

सोमपा नाम विप्राणां क्षत्रियाणां हविर्भुजः।
वैपानाभाज्यपा नाम शूद्राणां तु सुकालिनः।। 8,198

*That is, Sompa, Havirbhuj, Ajyapa and Sukalin are the names of the ancestors of Brahmans, Kshtriys, Vaishys and Shudrs respectively.*

सोमपास्तु कवेः पुत्राः हविष्मतोऽङ्गिरस्सुताः।
पुलस्त्यस्याज्यपाः पुत्राः वसिष्ठस्य सुकालिनः।। 8.199

*Sompa is the son of Bhrigu, Havirbhuj is the son of Angiras, Ajyapa is the son of Pulastya, Sukalin is the son of Vashishth. According to this lineage, Shudrs are the sons of Sukalin, Sukalin is the son of Vashishth, Vashishth is the son of Bhagwan Manu.*

*Brahmans*, *Kshtriys*, *Vaishys* and *Shudrs* all are born from Bhagwan Manu. With this the *Manusmriti* also propounds the principle of one lineage-one father. Except for the projected verses of all the *Smritis* like the *Yajnvalky Smriti*, *Manusmriti*, *Naradsmriti* etc., the remaining principles are very exemplary for social life and are very useful and excellent even today. In fact, *Shruti*-era would be considered as an ideal period. These discriminatory verses seem to be the work of the sponsored miscreants. Exposing them is the *Dharm* in the present times.

Later on, *Brahman* texts came to an existence, mainly in these *Brahman* texts, rituals are described. These texts are said to be of the priestly tradition. In this era, mainly controversial and discriminatory *Sutrs* were composed. At that time an Islamic state had already been established in the country.

# References from Mahabharat

*Varn System* based on birth has been negated. *Shlok* no. 22 of chapter 13 of the *Anushasan Parv* of the review edition of the *Mahabharat*, published by the Bhandarkar Research Institute, Pune, in the appendix of Kumbh Konam, Madras edition and Vadiraj's commentary as in appendix after *Shlok* 22, on this point are available. The *Mahabharat* has a clear concept that the clans of the *Rishis* and origins of the rivers are not to be questioned. *Rishis* attained *Brahmantav* with great austerities.

ऋषीणां च नदीनां च साधूनां च महात्मनाम।
प्रभवो नाधिगन्तव्यः स्त्रीणां दुश्चरितस्य च।।

A study of the *Mahabharat*, without prejudice, reveals what is a liberal and very broad-based social system. Satyavati, the daughter of *Nishad*, became the wife of the *Kshtriy* Maharaja Shantanu. Vyas, the child of Satyavati, before her marriage, fathered through *Niyog,* the son of both Amba and Ambalika and the servant's son Vidur. Vidur got a high position in the *Rajya Sabha*, and Bhagwan Shri Krishn visited his house. Society gave utmost respect to Arjun's marriage to *Nag Kanya* Ulupi, Bhim's marriage to the demon girl Hidamba, marriage of Jarasandh's sister Rukmini to Bhagwan Shri Krishn. Draupadi, even after becoming the wife of five husbands is called *Sati* (Chaste and Pious). These are clear examples of broadness of *Bhartiy* thinking.

78[th] chapter of the *Shantiparv* of the *Mahabharat* clearly mentions that a *Shudr* can also become a king and is entitled to all royal honours. The important dialogue between Bhishm Pitamah and Yudhishthir regarding *Rajdharm* is noteworthy in this context. Pitamah said that according to spacio-temporal reasons and other circumstances, *Dharm* becomes *Adharm.* Sometimes *Adharm* becomes *Dharm*.

Yudhishthir asked another question to Bhishm Pitamah : if the courge of dacoits increases in the society, the abduction of women leads to an increase in *Varn*-hybridity or people of all *Varn* get bewildered, how to protect the subjects. Can any capable *Brahman*, *Vaishy* or *Shudr* have the sanctity? Can he righteously perform the *Rajdharm* i.e. can he rule the kingdom? Should he perform all duties of King or not? Should he be stopped from doing such duties?

I am of the opinion that on such occasions people belonging to different *Varns* can also take up arms.

In reply to Yudhishthir's question, Bhishm Pitamah said :

अपारे यो भवेत् पारमलप्लवे यः प्लवो भवेत्।
शूद्रो वा यदिवाप्यन्य सर्वथा मानमर्हति। शान्तिपर्व 78,38

*O Dharmaraja Yudhishthir! The one who saves others from great distress and who, like a board provides support to drowning people, may he be a Shudr or anyone else, is worthy of respect.*

There is neither a sense of untouchability in the above *Shlok* of the *Mahabharat*, nor being a *Shudr* a disqualification to sit on the throne. In other *Shloks* of *Shantiparv* in 78th chapter, Pitamah has considered the capability and not the *Varn* as a criteria for a king. In 44th *Shlok* of this chapter he further said:

नित्यं यस्तु सतो रक्षेदजतश्च निदर्तयोत्।
स एव राजा कर्तव्यस्तेन सर्वमिदंधृतम्।। शान्तिपर्व

That is, the one who always protects the good men and suppresses the wicked should be made the king because through him, the whole kingdom remains safe. This means that to be a king, it is not necessary to be a *Kshtriy* or a *Brahman*. *Shudr* is equally entitled to become king. Whoever, may be from any *Varn*, has the power to punish the wicked and keep the gentle people safe can become a king.

The sermons given in the chapter *Shantiparv* by Bhishm Pitamah to Yudhishthir regarding the formation of council of ministers are very appropriate and relevant on the subject :

चतुरो ब्राह्मणान् वैद्यान प्रगल्भान स्नातकान् शुचीन।
त्रींश्च शूद्रान विनीतांश्च शुचीन कर्मणि पूर्व के।।
अष्टाभिश्चगुणैर्युक्तं सूतं पौराणिक तथा।
पञ्चाशद्वर्षवयसं प्रगल्भमनसूयकम्।।
श्रुतिस्मृति समायुक्तं विनीतं समदर्शनम्।

कार्ये विबदमानानां सक्तमर्थेश्वलोलुपम्।।
वर्जित चैक्यव्यसनैः सुघोरैः सप्तभिर्भृशम्।
अष्टानां मन्त्रिषां मध्ये मन्त्रं राजोपधारयेत।। 85,7–10

*In the council of eight ministers, the king should place four Brahman Vedic scholars who are fearless, virtuous and hold academic degrees. Three pious, devout Shudrs who embody fundamental values. One should be a Sut (Charioteer) having eight virtues. After consulting these eight ministers the king should run the kingdom. These people should be around fifty years of age and should be fearless, free from jealousy, greed and addiction; endowed with knowledge of Veds and Smritis; should be gracious, sympathetic, capable of judicial decisions. (85 7, 8, 9, 10)*

The above *Shlok* is available in all the additions of the *Mahabharat*. Astu! It is clear from this that *Shudrs* were respected as ministers in the council of ministers of kings. Three out of eight ministers were *Shudr*. This gives insight in to the mind of the *Bhartiy Rishis*.

A few people, having anti-Hindu mentality propagate that *Shudrs* do not have the right to perform *Yagy*, whereas chapter 60 *Shlok* 39 of the chapter *Shantiparv* of the *Mahabharat* is enough to dispel this notion.

शूद्रः पैजवनो नाम सहस्त्राणां शतं ददौ।
ऐन्द्राग्नेन विधानेन दक्षिणामिति नः श्रुतम्।।

*Pitamah Bhishm said to Yudhishthir that we have heard that in ancient times a Shudr named Paijavan had performed a ritualistic sacrifice by invoking Indr and Agni and donated one lakh filled pots as Dakshina. (gift)*

Dr. Ambedkar has also called this *Shlok* revolutionary.

अहो हि सर्व वर्णानां श्रद्धायज्ञो विधीयते।
देवतं हि महच्छ्रद्धा पवित्रं यजतां च यत्।। शान्तिपर्व 61,40

*Therefore, there is a law for all Varns to perform Yagy with devotion. Devotion to the great deity purifies the performer of Yagy.*

There is a dialogue of Narad *Muni* and *Rishi* Markandeya in the 54th chapter of *Shantiparv* of the *Mahabharat*, in which *Rishi* Markandeya questions Narad *Muni*:

चत्वारो ह्यथ ये वर्णा हव्यकव्य प्रदास्यते।
मन्त्रहीनमपन्यायं तेषां दत्तं क्व गच्छति।।

*That is, O Narad ji, what happens to the Hvya-Kvya offered by the people of the four Varn, without chanting and following the protocol ? Narad ji answered the Rishi :*

असुरान् गच्छते दत्तं, विप्रैः रक्षांसि क्षत्रियैः।
वैश्यैः प्रेतानि वै दत्तं शूद्रैर्भूतानि गच्छति।।

*Such Yagy offerings called Hvya offered by the Vipras goes to devils, the Kshtriys to the demons, the Vaishys to the ghosts, and the Shudrs to the ghouls.*

Offerings of *Anna Pind* given to the ancestors in *Shradh* is called *Kvya*. It is clear that along with other *Varns*, the *Shudrs* were not only allowed to recite *Mantrs*, perform *Yagy*, but were also allowed to perform rituals and had equal rights like any other *Varn*.

There is a description of the creation of Universe and the creation of *Varns* in the chapters 188 and 189 of *Shantiparv* of the *Mahabharat*. In these chapters concepts given in the *Purushsukt*, the *Mundokupanishad*, the *Brihadaranyakupanishad* about the creation of Universe are accepted. In these chapters according to the *Shloks* it is believed that in ancient times all human beings were *Brahmans*. Other *Varns* came into existence at a later stage but had equal rights to *Vedavani* and *Yagy*. The basis of *Varns* is not birth but *Karm*. If a *Brahman* does not have the ascribed virtues to be a *Brahman*, then he is a *Shudr* and a *Shudr* who has the ascribed virtue will be considered as a *Brahman*.

*Maharishi* Bhrigu said to Bharadwaj *Muni* :

असृजद ब्राह्मणानेव पूर्व ब्रह्मा प्रजापतिः।
आत्म तेजोभिनिर्वृत्तान् भास्कराग्निसमप्रभान।।
ब्राह्मणाः, क्षत्रिया, वैश्याः शूद्राश्च, द्विज सत्तम।
ये चान्ये भूत सङ्धानाम् सङ्धास्तांश्चापि निर्ममें।।

That is, at the beginning of the creation, Prajapati *Brahm*, with his brilliance, created only *Brahmans* who were as luminous as the sun and fire. After that, as a means of attaining heaven, Prajapati *Brahm* gave the law of truth, *Dharm*, austerity, *Sanatan Veds*, ethics and purity. After this *Brahm* created Gods, devil, Gandharvas, Daityas, Asuras, big snakes, demons, yakshas, vipras, vampires and humans subsequently he created *Brahmans*, *Kshtriys*, *Vaishys*, *Shudrs* and also created other species of breathing beings.

Bhrigu *Rishi* further said :

न विशेशोऽस्ति वर्णानाम् सर्व ब्राह्ममिदं जगत्।
ब्रह्मणा पूर्व सृष्टं यत् कर्मभिर्वर्णतां गतम्।।

*In ancient times there was no special Varn. As the Universe was created by Brahm, the world itself was Brahmmaya or Brahman. Later, on the basis of Karm, Varns came in to existence.*

It has been clarified that *Varns* are identified on the basis of desired attributes and *Karm*. A person in whom the desired attributes and *Karm* of the specific *Varn* are visible then that person will be considered to be of that *Varn*. The virtues of *Brahmans*, *Kshtriys*, *Vaishys* and *Shudrs* have been defined by the *Rishi* Bhrigu as follows :

सत्यं दानमथाद्रोहं आनृशंस्यं त्रपा घृणा।
तपश्च दृश्यते यत्र स ब्राह्मण इति स्मृतः।।

*That is, in whom attributes of the spirit of truth, charity, loyalty, non-violence, forgiveness, mercy, austerity are seen then he is considered a Brahman.*

क्षत्रजं सेवते कर्म वेदाध्ययनसंगतः।
दानादानरतिर्यस्तु स वै क्षत्रिय उच्यते।।

*One who is involved in deeds like war etc., is busy in the study of Veds, gives charity to eligible persons and makes the subjects of his kingdom obey the law is called Kshtriy.*

वाणिज्या पशुरक्षा च कृष्यादानरतिः शुचिः।
वेदाध्ययनसम्पन्नः स वैश्य इति संज्ञितः।।

*A person, who is engaged in the study of Veds even after being engaged in trade, animal husbandry and agricultural work is called Vaishy.*

सर्वभक्षरतिर्नित्यं सर्वकर्मकरोऽशुचिः।
यक्तवेदस्त्वनाचारः स वै शूद्र इति स्मृतः।।

*That person who eats anything without considering its health benefits; is engaged in all kinds of degraded, condemned or blasphemed activities; who has discarded the study of Veds is impure and immoral and is considered a Shudr.*

In the *Shlok* being quoted after the above *Shloks*, there is a confirmed declaration regarding how a change of *Varn* from one to another *Varn* takes place.

शूद्रे चैतद् भवेल्लक्ष्यं द्विजे तच्च न विद्यते।
न वै शूद्रो भवेच्छूद्रो ब्राह्मणों न च ब्राह्मणः।।

*That is, if the signs of virtue like truth, charity etc. are seen in a person, then he is not a Shudr and if a person does not have these attributes of virtue, then he is not a Brahman.*

The above *Shlok* is sufficient to prove that the *Varn System* is not based on birth but on the basis of distinctive attributes and *Karm*. A *Brahman*, without virtues, ceases to be a *Brahman* and becomes a *Shudr*, and a person born in a *Shudr* family but has these attributes becomes a *Brahman*.

In the *Udyogparv* of the *Mahabharat*, a conversation is described between Yudhishthir and King Nahush who had become a python. In this conversation, in relation to the *Varn System* a similar theory has been propounded. Yudhishthir's answer to the question of Python (King Nahush) is as follows :

शूद्रे तु यद् भवेल्लक्ष्य द्विजे तच्च न विद्यते।
न वै शूद्रो भवेच्छूद्रो ब्राह्मणों न च ब्राह्मणः।।
यत्रैतल्लभ्यते सर्प वृत्तं स ब्राह्मणः स्मृतः।
यत्रैतत्र भवेत् सर्प तं शूद्रमिति निर्दिशेत्।।

*Yudhishthir said, O Python! If these attributes (truth and charity) are present in a Shudr that Shudr is not a Shudr, and if these are not present in a Brahman then he is not a Brahman. Any person who shows the signs of truth, that person is a Brahman and who does not have these attributes, he may be a Brahman by birth but in fact he is a Shudr.*

A *Shlok* in the similar context is very important in this regard :

तावच्छूद्रसमो ह्येश यावद वेदे न जायते।
तस्मित्रेव मनिद्वैधे मनुः स्वायम्भुवोऽब्रवीत।

*That is, until the child is not initiated in the Veds, that is after Sanskars child is not taught Veds till then that child remains like a Shudr. This is what Swayambhu Manu said:*

The Yaksh-Yudhisthir dialogue described in the 313th chapter of the *Vanparv* of the *Mahabharat* is very remarkable in relation to the *Varn System*. In fact, Yudhishthir's father, in the form of Yaksh wanted to test his son Yudhishthir's intelligence, scholarship and eloquence. Yudhishthir in a decisive voice gave the same answers to Yaksh which he had given to Nahush (Python) in *Udyogparv*.

चतुर्वेदोऽपि दुर्वृत्तः स शूद्रादतिरिच्यते।
योऽग्निहोत्रपरो दान्तः स ब्राह्मण इति स्मृतः।।

*According to Yudhishthir, if a scholar of the four Veds is a misdemeanor then he is a Shudr, and if a person who performs Agnihotr and is Stoic, then he is a Brahman.*

According to this, the basis for determination of the *Varn* is the ethical behaviour and not clan or erudition.

There is an anecdote in the 206th chapter of *Vanparv* of the *Mahabharat* of Dharmvyadh preaching *Dharm* to Kaushik, a *Brahman*. It is evident from this anecdote that the right of sermon was available to every person of every *Varn*, whose conduct was pure and chaste. A similar anecdote is found in the 260th chapter of *Shantiparv* in which there is Tuladhar, a *Vaishy* preaches *Dharm* to a great scholar like Jajli *Muni*.

A very interesting incident in the 10th chapter of *Anushasan Parv* of the *Mahabharat* throws light on the social conditions of that time. According to this incident, Bhishm Pitamah narrates a story to Yudhishthir. Once a pious *Shudr* went to the hermitage of ascetic *Brahmans* on the holy Himalayan mountain. The ascetics of the *Ashram* treated the *Shudr* with full respect. Seeing the Sannyasis, there was a desire in the mind of that *Shudr* to become a *Sanyasi*. The Vice-Chancellor of the *Ashram* gave him permission to stay there, but refused to initiate him into *Sannyas*. On this, the determined *Shudr* built a hut some distance away from the *Ashram*. He also built a *Yagyvedi* and temple and started living like a *Muni* (follower of rules). Gradually, that *Shudr Muni* was recognized and respected amongst the *Brahman Rishis*. After the death and his merging in the *Panchatattv* that *Shudr Muni*, because of his pious and good deeds, was born in a royal dynasty. He earned prestige by becoming a Bhupati. At the same time, the *Brahman Rishi* who had earlier performed the *Shradh* ceremony of that *Shudr* Muni's father was appointed as *Rajpurohit*. On seeing *Rajpurohit*, reciting holy *Mantrs* and performing rituals Bhupati with a twinkle smiled at *Rajpurohit,* which he did not like. When the *Rajpurohit* asked the reason for this twinkling smile, the Bhupati narrated the entire story of his and that of *Rajpurohit's* previous birth.

After a complete retrospective overview of the *Mahabharat*, we can conclude the following :

1. *Varn* is not determined on the basis of birth but on the basis of attributes and *Karm i.e. Gun-Dharm.*
2. If true virtues are there at the time of birth in a *Shudr*, then he will be considered a *Brahman*, and if there is a lack of these virtues then that person will be considered a *Shudr*.
3. In the *Dwaparyug*, the *Varn* hybridization was very strong and children by cross-breedings were born in all the *Varns*; it is difficult to determine the *Varn* on the criteria of birth. The claim of *Varn Shuddhi* is false and unnecessary in the present times.
4. As the basic instincts of people of all *Varns* are same, there cannot be any discrimination on the basis of birth.
5. On the basis of *Karm* and *Sanskar*, mother of every child from every *Varn* is called Savitri and father is called Achary. Therefore, there cannot be any discrimination on the basis of birth.
6. As long as a person does not study the *Veds*, he is considered a *Shudr*. A person born as a *Brahman* but who does not study the *Ved* is not a *Brahman*.
7. For distinctions among the *Varns*, *Rishis* and seers have given utmost significance to character and not to family or birth.
8. Neither austerity nor sacrificial rituals were prohibited for *Shudrs* and females.
9. On the basis of all these findings, it is quite clear that our *Varn* determination was done not on the basis of birth, but on the basis of distinctive attributes and *Karm*.

## References from Gita

There is no complex problem of human life for which solution is not provided in the *Gita* narrated by Bhagwan Shri Krishn. Scholars are of the opinion that educated people of all the countries, of all languages, of all castes, have perused the *Gita*. Bal Gangadhar Tilak, thinker of Bharat, in his book the *Shrimad Bhagwad Gita Rahasya* has called the *Shrimad Bhagwad Gita* a stunning and a pure diamond amongst our *Dharmik* texts. It explains in a short and clear way, the esoteric and sacred elements of self-knowledge including the *Pind-Brahm-Paravidya* (body, universe and the ultimate knowledge). Based on these elements, the *Gita* introduces the concept of *Purusharth* to human beings through which perfect physical, cosmic and ultimate spiritual consciousness can be attained. It combines devotion with knowledge and brings it under the umbrella of scriptural conduct. It propels every human being struck by worldly griefs towards performance of duties with *Nishkam Bhaav* (selfless spirit). Other than the *Gita* there is no illuminating, comprehensive and perceptive spiritual texts in the world literature, what to say in Sanskrit language. Since the *Gita* emanated from lotus-mouth of *Parabrahm Parmeshwar Yogeshwar* Bhagwan Shri Krishn, it is considered to be a holy and a divine expression. According to scholars, after experientially perusing the *Gita*, there is no need to study any other scripture. The *Gita* is considered to be the *Triveni* (merging of three rivers) of knowledge, devotion and action. Dr. Radha Krishnan in his book the *Bhagwatam Gita* has quoted the following lines of Aldous Huxley in this regard, "The *Gita* is one of the clearest and the most comprehensive summaries of the perennial philosophy ever to have been made. Hence, it has enduring value, not only for *Bhartiys*, but for all mankind...... The *Bhagwad Gita* is perhaps the most systematic spiritual statement of the perennial philosophy."

This implies that the *Gita* has been proclaimed as the most crystallized and comprehensive of the *Sanatan* philosophical texts. Therefore, its essence has a lasting impact not only on every *Bhartiy*, but also on mankind as a whole. The *Bhagwad Gita* is perhaps the most well-organized spiritual statement of *Sanatan Darshan*. Thus, the *Gita* as a *Dharmik* text has a far reaching impression and indelible influence on subsequent schools of thought and philosophy.

The precepts of the *Gita* are very useful, expedient and relevant even on the subject of *Bhartiy Varn System*. In fact, the vision of the *Gita* is all pervading in relation to the entire creation:

ब्रहमभूतः प्रसन्नात्मा न शोचति न काङ्क्षति।
समः सर्वेषु भूतेषु मद्भक्तिं लभते पराम्।। गीता 18,54

*By becoming engrossed in Brahm - a calm soul, neither lamenting nor craving and beholding equality in all begins-gains supreme devotion toward Me. (Gita 18,54)*

According to the *Gita*, being always engaged in the welfare of all beings is also an essential part of the devotion to Supreme power. The *Gita* defines this quality with the title of 'सर्वभूतहितरेतः' On the basis of this equitable principle, Bhagwan Shri Krishn has presented a comprehensive view of the *Varn System* in the *Gita*. This statement of Bhagwan Shri Krishn in relation to the *Varn System* is the foundation stone of *Sanatan Dharm* :

चातुर्वर्ण्यं मया सृष्टं गुणकर्मविभागशः।
तस्य कर्तारमपि मां विद्धयकर्तारमव्ययम्।। गीता 4,13

*According to the differentiation of attributes and Karm, I have created the four Varns. Though, I am the Doer, yet know Me to be Non-performer, beyond all change. (Gita 4,13)*

The basis of the *Varn System* mentioned in the *Gita* is not by birth. The Bhagwan has defined attributes and *Karm* to be the basis of the division of *Varn*. Bhagwan Shri Krishn gives a detailed explanation of the *Varn System* in :

ब्राह्मणक्षत्रियविशां शूद्राणां च परन्तप।
कर्माणि प्रविभक्तानि स्वाभावप्रभवैर्गुणैः।। गीता 18,41

*O Scorcher of Foes! The duties of Brahmans, of Kshtriys, of*

*Vaishys, as also of Shudrs, are allocated according to the attributes springing from their own nature. (Gita 18, 41)*

In fact, the determination of *Karm* on the basis of inherent nature and latent tendencies is the best scientific method. Political scholars and academicians from all over the world have sought it.

Experts opine that according to the latent tendencies of the child, opportunity should be provided to children for the ripening of the latent tendencies in them. They also concur that through psychological investigations, it must be identified as to which qualities i.e. of a poet, writer, engineer, doctor, craftsman or businessman are present in the child. The education-initiation of the child should be determined on the basis of qualities thus identified. Once Swami Vivekanand expressed the view, "Manifestation of perfection already in child is the object of education". This was also expressed in a program of the school run by sister Nivedita.

The aim of education is to increase the inherent attributes in the child qualitatively. In the opinion of politicians the concept of idealist state can be realized only if the child is developed based on its latent tendencies. Therefore, the principle of division of the *Varn* on the basis distinctive attributes and *Karm* is completely scientific and nourishes the goal of an Ideal State. Bhagwan Shri Krishn referred to natural qualities :

शमो दमस्तपः शौचं शान्तिरार्जवमेव च।
ज्ञानं विज्ञानमास्तिक्यं ब्रह्मकर्म स्वभावजम्।। 18,42

*Mental-control, sense-control, self-discipline, purity, forgiveness, honesty, wisdom, self-realization, and the faith in rebirth, constitute the duties of Brahmans, springing from their own nature. (18,42)*

शौर्यं तेजो धृतिर्दाक्ष्यं युद्धे चाप्यपलायनम्।
दानमीश्वरभावश्च क्षात्रं कर्म स्वभावजम्।। 18,43

*Valor, radiance, resolute endurance, skillfulness, not fleeing from battle, beneficence and leadership are the natural duties of the Kshtriys. (18,43)*

कृषिगौरक्ष्य वाणिज्यं वैश्यकर्म स्वभावजम्।
परिचर्यात्मकं कर्म शूद्रस्यापि स्वभावजनम्।। 18,44

*Tilling the soil, cattle breeding and business are the natural duties of the Vaishys. Actions that are to serve the others are natural duty of the Shudrs. (18,44)*

An incisive study of the above *Shloks* makes it evident that the inherent qualities of *Brahman*, *Kshtriy*, *Vaishy*, *Shudr* are present in every child at the time of birth. Through a systematic effort at development of these innate natural qualities, a child can attain specialization in a particular field. Bhagwan Shri Krishn said that only a person engaged in work according to his attributes attains the ultimate accomplishment:

स्वे स्वे कर्मण्यभिरतः संसिद्धिं लभते नरः।
स्वकर्मनिरतः सिद्धिं यथा विन्दति तच्छृणु।। 18,45

*With attentiveness to his own duty, man gains the highest success. Through devotion to his inborn duty, he attains success. (18,45)*

यतः प्रवृत्तिभूतानां येन सर्वमिदं ततम्।
स्वकर्मणा तमभ्यर्च्य सिद्धिं विन्दति मानवः। 18,46

*A man attains perfection by worshiping, with his natural gifts, Him from whom all being are evolved, and with whom all this world is permeated.(18,46)*

The meaning is that this entire creation is a manifestation of Bhagwan. So the entire creation is His form. When a person serves this creation on the basis of his natural attributes, then in a way he serves Bhagwan. According to the principle of 'ईषावास्यमिंद सर्वम' (Ishavasyamidam Sarvam), worshiping the universe as *Janardan* (common masses) means worshipping Bhagwan. A man does not commit sin by doing the work according to his attributes.

The four *Varn System*, in fact, was prescribed for the development of values and the orderly operation of the society. Seeing the natural actions of *Brahmans*, *Kshtriys* and *Vaishys*, the *Shudrs* should not have any inferiority complex and should not give up their work thinking that their work is inferior to the other three *Varns*. Bhagwan Shri Krishn preached to serve society according to one's own *Dharm* :

श्रेयान्स्वधर्मो विगुणः परधर्मात्स्वनुष्ठितात्।
स्वभावनियतं कर्म कुर्वन्नाप्नोति किल्विषम्। 18,47

सहजं कर्म कौन्तेय सदोषमपि न त्यजेत्।
सर्वारम्भा हि दोषेण धूमेनाग्निरिवावृताः ।। 18,48

*Better than the well-accomplished Dharm of another is one's own Dharm, even though lacking merit (somewhat imperfect). He who performs the duty decreed by his inborn nature contracts no sin. (18,47)*

*O Offspring of Kunti (Arjun), one should not abandon one' inborn duty, even though it has some imperfections, for all undertakings are marred by blemishes, as flame by smoke. (18,48)*

How simple is the explanation of actions based on natural attributes and spontaneity. The concept of *Karm* determination and *Varn* division is a refined scientific concept. This scientific concept of *Varn System* of the *Gita* is a manifestation of the fundamental spirit of *Sanatan Dharm.*

Those who attribute *Varn System* to birth should read the *Varn System* as explained by Bhagwan Shri Krishn in the *Gita* and understand it with the eyes of discernment, renouncing perversity. With this, along with the welfare of the *Sanatan* Hindu society, the welfare of the entire world is possible because only *Sanatan Dharm* in the world is the human *Dharm* and is above any religion, creed, and sect.

# References from Ramayan

*Bhartiy* scholars believe that the *Ramayan* narrated by *Maharishi* Valmiki is the basis of the *Ramcharitmanas*. They also believe that through *Rishi* Valmiki, the four *Veds* appeared in the form of *Ramayan* at the time when *Parabrahm Parmeshwar* incarnated as Ram in the house of King Dasharath of Ayodhya. In fact, *Maharishi* Valmiki is considered to be the *Adi Kavi* (Primordial Poet) of India and the *Ramayan* composed by him is *Adi Granth* (Primordial *Granth*). Till date the *Ramayan* is revered as a sublime *Granth*. Since antiquity the *Ramayan* has been the focal point of literary creativity.

*Bhartiy* scholars also believe that once upon a time with the intention of plunder, a dacoit called Ratnakar imprisoned *Saptarishis*. A dialogue initiated by *Saptarishis* transformed Ratnakar into a sensitive, ethical and moral being. Witnessing the death of one partner of the crain couple, poetry erupted from the empathetic heart of Ratnakar and he became *Maharishi* Valmiki.

In the context of social harmony and *Varn System*, *Valmiki Ramayan* is also seen presenting the sublime form of the fundamental concept of *Bhartiy* ancient texts. The "Balkand" of the *Ramayan*, gives permission to all *Varns* to recite the *Ramayan*:

इदं पवित्र पापघ्नं पुण्यं वेदैश्च सम्मितम् ।
यः पठेद् रामचरितम् सर्वपापैः प्रमुच्यते ।। 1,97

*That is, one who recites the Ramayan, which was considered like the Veds, will be free from sins; as it is holy, sin-destroyer and pious like the Veds. (1,97)*

पठन द्विजो वागृषभत्वमीयात् । स्यात् क्षत्रियो भूमिपतित्वमीयात् ।।
वणिग्जनः पुण्यफलत्वमीयाज्–जनश्च शूद्रोऽपि महत्वमीयात् ।। 1,100

*The meaning is that everyone has the right to read the Ramayan, if a Brahman reads it, then he gets proficiency in speech, if a Kshtriy reads it, then he becomes a king on earth, if a Vaishy reads, then his business becomes more profitable and if a Shudr reads he gets importance in society.*

Similar views are expressed in the *Bhagavatam* also, which have been cited earlier. Apart from this, many episodes of the Valmiki *Ramayan* are filled with the auspicious feeling of सर्वजनहिताय, सर्वजनसुखाय (welfare of everyone, happiness of everyone). Similarly, these principles are also expounded in *Ramcharitmanas*. In fact, Tulsidas is considered to be an incarnation of Valmiki. Saint Nabhadas said, "कलि, कुटिल, जीव निस्तारहित, वाल्मीकि तुलसी भयो" (In *Kalyug*, for the betterment of people, Valmiki, was born as Tulsi). The references of the *Ramcharitmanas* are also for public welfare. The followers of Hindu *Dharm* says that it is a unique and ultimate *Granth* with extracts of principles enshrined in the *Ved*, *Puran* and *Upanishads*. At the very beginning of the composition of the *Ramcharitmanas*, Tulsi makes it clear that the *Ramcharitmanas* is compatible with many *Purans*, *Veds* and scriptures. The *Ramcharitmanas* is the essence of the ideal principles of social life and values as propounded in the various Hindu scriptures. In the *Ramcharitmanas*, every aspects of human life is considered and contemplated upon in detail. So is the case with the *Varn System*. Clarifying the concept of the *Varn System* Tulsidas said:

वरनाश्रम निज निज धरम, निरत वेद पथ लोग।
चलहिं सदा पावहिं सुखहिं, नहिं भय सोक न रोग। उ० का० 20

*All the people, ready to follow the Dharm according to their respective Varn and Ashram, always walk on the path of the Veds and find happiness. They have no fear, grief or disease and misfortune does not bother them. (Uttar Kand 20)*

This description of *Ramrajy* is the proof that *Varn System* was thoughtfully created for social harmony and progress. *Varn System* is its key to solve prevailing problems. Perhaps it is the only solution. Tulsi discussed the fall of this system and said:

वरन धरम नहिं आश्रम चारी।
श्रुति विरोध रत सब नर नारी।। उ० का० 97

*In Kali Yuga, neither Varn Dharm exists nor the four Ashrams. All men and women will be engaged in opposing the Veds. This description of Kaliyug is very effective and based on reality.*

Tulsi further clarified :

एक पिता के विपुल कुमारा। होंहिं पृथक गुन शील अचारा।।
कोउ पण्डित कोउ तापस ज्ञाता। कोउ धनवंत सूर कोउ दाता।।

*Tendencies and qualities of sons of one father may be different. One son may be learned, other son may be an ascetic, the other may be wealthy, brave and charitable.*

Similarly, according to inherent qualities, a person determines his path by performing *Karm* and attains his *Varn*. The father loves all children equally. Similarly, only a visionary approach for the deprived, exploited and weak will re-establish the social welfare system. Any movement in opposite direction will weaken the Hindu Society.

The essence of the entire *Ramcharitmanas* is visible in *Uttarkand*. The high peaks of good qualities of ancient times and the low ebbs of modern times are visible in it. Surprisingly, some *Chaupais* also appear to be incompatible with each other. Like in other scriptures, these are malicious interpolations. If we take into consideration the period of *Ramrajy*, then we have unique proclamations of social harmony. Consider this *Chaupai* describing the effects of Ram's coronation :

राम राज बैठे त्रैलोका। हर्षित भये गये सब शोका।।

*When Shri Ram Chandr became king, all the three worlds were happy; sorrows disappeared.*

Ram presented himself as *Maryada Purushottam* throughout his whole life. Nowhere does he seem to be a miracle-man. But the description given in the *Chaupais* is no less than a miracle. The *Chaupa is* following the above also describe similar miracles :

*As soon as Ram's coronation takes place, no one hates anyone. With the glory of Shri Ram, everyone's disparity vanishes. Tritap (Physical, Psychological and illnesses) do not affect anyone. All follow their respective Dharm by following the policy stated in the Veds. By following the four steps i.e. Truth, Purity, Mercy and Charity, Dharm gets repleted and perfected in the world. There is no sin. Men and women are all devoted to Ram Bhakti and all are entitled to Moksh. Neither anyone has short life-span nor does anyone suffer pain. Everyone's body is beautiful and healthy. No one is poor, sad, worried, plaintive, foolish or devoid of good qualities. All are conscientious following Dharm and are pious and saintly. Men and women are all astute and virtuous and respect the learned and knowledgeable people. All are grateful. There is no deceit and cunning in anyone.*

This form of ideal social formation and excellent state management is distinctive and timeless in itself. All modern scholars including Mahatma Gandhi have considered this exemplary system of state and society to be not only relevant, but indispensable and meritorious. In this system, the common interest of all *Varns* are taken care of. As per the *Ramcharitmanas*, this too is the *Vedic* system. One more thing that strikes to my mind, which I have described in preceeding pages, is how could such a miracle happen as soon as Ram's coronation took place that everything became ideal. Ram, in his entire life never performed any miracle. Ram had tried to guide the whole world through his character and conduct as to how an ordinary man can lead a superior and sublime life by maintaining his *Maryadit* life. Then how could this miracle happen? In this context, I am of the belief that Ram did not perform any miracle, nor did it happen because of his coronation. It was the result of *Maryadit* life of forest dwellers led by Ram along with his younger brother Lakshman as well as because of the indefatigable efforts of Bharat and Shatrughn that *Ramrajy* got established. To understand the new social harmony of that time, we have to seriously examine the 14 years of his life as a forest dweller.

There is a context in the *Aranyakand* that as soon as Ram went to forest, he met many *Rishis*. During this journey, at one of the places, he saw a pile of bones. 'अस्थि समूह देखि रघुराया। पूछी मुनिन्ह लागि अति दाया।।' Rishis informed Ram that these heaps of bones are of those Rishis who were killed by demons. Empathetic Ram was moved and was full of tears. In a determined way, Ram clenched his fists, raised his hands and took a resolve that : निसिचरहीन करउँ महि, भुज उठाय पन कीन्ह।। (Ram resoled that I will make this earth demon free.)

Just think! Vanvasi (forest dweller) Ram accompanied by his wife Sita and younger brother Lakshman had neither weapons, nor means, or army. How could this resolve be fulfilled?

The efforts made by Ram in fulfilling this resolve, became desideratum in the establishment of the *Ramrajy* after his coronation. This innovative, ideal, social creation and management of the state system was visible to the general public.

Ram had set a goal for 14 years of exile 'to make the whole earth demon-free.' For this he made a wonderful plan. It is clear that he needed weapons, army and means to achieve his goals. So he started working on his plan immediately after the resolve. In the above *Doha* (couplet) of *Aranykand*, in which Ram resolved to make the earth demon-free. Tulsi reveals a part of his plan, saying that, सकल मुनिन्ह के आश्रमहिं जाइ जाइ सुख दीन्ह।। *(He went to the Ashrams of Rishis to seek their cooperation and blessings)*.

In *Rishi* tradition of Bharat, scientists with spiritual excellence were involved. The word 'Researcher' of English is derived from the word *Rishi*. Ram obtained weapons from these *Rishis* (scientists). To form his army, he embraced and involved the exploited, oppressed, deprived, *Girivasis* (mountain dwellers), *Adivasis, Kol, Bhil, Kirat* (various tribes of India) of the society. To spread the message of mutual love and to eliminate the inequality prevalent in society, Ram ate jujube berries of Shabri. He killed the wicked and infused courage in society by creating fearlessness. As a result of this infusion, those very people who were trembling with the fear of the demons, killed the great mighty evil emperor Ravan, wiped out the entire *Rakshas* (demon) culture. *Ramrajy* is the result of this innovative effort of 14 years.

A look at a few *Chaupais* from the *Ramcharitmanas* reveals the solution for today's castes problems. According to the *Vedic* system, mutual love can exist even if the worshipping methods are different. A perceptive and human approach as proclaimed by the *Veds* is necessary for the achieving coexistence and collectiveness. A constricted thinking cannot be *Vedic* thinking. Today, society is placed at verge of an explosive volcano of mutual animosity. Several anti-social elements are working to fuel the fire. Hindu contemplation, which The Honorable Supreme Court of India has hailed in one of its decisions says that a broad and liberal thinking equal to Hindu contemplations may not be found anywhere else in the world.

If we make this innovative *Mantr* of the *Ramcharitmanas* i.e. the resolve of Ram a part of our life, India will be able to regain its prestige. This distinctive principle of social harmony can act as the backbone of the society. There will be no disparity if we implement Ram's ideal thinking and plan in modern times. When the disparities and inequalities vanish, everyone will love each other. The social harmony will be restored automatically. With the roots of Islamic terrorism and the accelerated pace of Christian religious conversions Bharat will be eliminated completely.

Inclusive vision of Ram as explained in the *Ramcharitmanas* is enough to demolish all the taboos imposed on the downtrodden, oppressed and deprived.

Out of selfishness and nefarious motives, as is the case with the other scriptures the *Ramcharitmanas* to has been tempered with; otherwise proclamations of the *Ramcharitmanas* are for the benefit of every living being. The need of the hour is to remove these interpolations and temperings and resolve to make the Ganga of the *Ramcharitmanas* pollution free.

The following *Chaupai* is constantly used as a weapon by Hindu-phobics against Hinduism.

ढोल गँवार शूद्र पशु नारी। सकल ताड़ना के अधिकारी।।

By interpreting the word *Shudr* on caste lines, it has been propagated that in the *Ramcharitmanas*, a *Shudr* has been described as a being deserving to be controlled and battered. It is completely untrue and misleading. For this reason Ramswami Periyar had the audacity to burn the copies or tear down the *Ramcharitmanas* at many places in Andhra Pradesh and many other states of southern Bharat. This is also a matter of concern that most of the group of preachers and *Sannyasis*, who keep preaching throughout the country, bowed their heads out of shame by accepting this misinterpretation. This would be treated as a bowed-down confession of the allegations leveled, and hence was further fuelled by Ramswami Periyar. As a result, a section of our own society became hostile to the *Ramcharitmanas*. If the leaders of the Hindu society had taken this campaign seriously and made strong protests by revealing the truth in a planned manner throughout the country, the impact of Periyar would not have been so widespread. These days a large section of the country and a political party are working on a project to make him a grandiose and respectable man. This is a serious challenge to the *Sanatan* society.

The spirit of the *Ramcharitmanas* is very comprehensive and oriented towards public welfare. Defining *Dharm* Tulsi said:

परमारथ अरु आतम पूजा। एहि ते धरम और नहिं दूजा।।
परहित सरिस धर्म नहिं भाई। पर पीड़ा सम नहिं अधमाई।।

*There is no greater Dharm than self-realization and self-contemplation. There is no Dharm greater than benevolence for the benefit of others. There is no greater unrighteousness than inflicting pain on others.*

Introspection and envisioning God in every being, as expressed in the Ishophanishad is reiterated in the following way by Tulsi, ईश्वर अंश जीव अविनाशी। चेतन अमल सहज सुखरासी। (Humans, as a part of Ishwar (God), are eternal, conscious, immaculate, spontaneous and fountain of happiness).

Along with the above mentioned *Chaupais*, the manifestation of the *Vedic* spirit of 'सर्व भूतहिते रतः' (keep doing good for all) permeates the entire *Ramcharitmanas*. Then how could the same Tulsi support the persecution of any section of the society? The Ocean, entity from whom these words were made to be uttered is neither a *Shudr* nor any other character mentioned in the above *Chaupai*.

In order to conquer Ravan, Ram, along with the army wanted to cross the sea. Vibhishan advised him :

प्रभु तुम्हार कुलगुर जलधि कहहि उपाय विचारि।
बिनु प्रयास सागर तरहिं सकल भालु कपि धारि।। सुन्दरकाण्ड 50

*'O Bhagwan! Samudr, the eldest ancestor in your clan, will devise and suggest a solution. The entire army of bears and monkeys will cross the sea effortlessly'.*

According to the context, after this advice by Vibhishan, Ram, for three continuous days kept pleading with the ocean to provide a solution to the problem of going across it. When the Ocean did not give way, Ram asked Lakshman to bring the bow and arrow. See the following *Chaupai* in this context and ponder over the following *Chaupai* :

मकर उरग झस गन अकुलाने। जरत जन्तु जलनिधि जब जाने।।
कनक थार भरि मनिगन नाना। विप्र रूप आयउ तजि माना।।

*The Ocean realized that with the anger of Ram all creatures living within him will be in trouble. Therefore, the Ocean, an ancestor of Ram, leaving his self-respect aside, appeared before Ram as a Vipr (Brahman). Exhibiting its superior scholarship the Ocean offers a golden platter full of rare gems to Ram.*

The word *'Vipr'* connotes a person of great scholarship. The word can also be used for *Brahmans*. Where is the *Shudr* here? I am unable to understand. One must appreciate the dialogue between Ram and the Ocean.

गगन समीर अनल जल धरनी। इनकइ नाथ सहज जड़ करनी।।
तव प्रेरित माया उपजाये। सृष्टि हेतु सब ग्रन्थन गाये।।
प्रभु आयसु जेहि कहँ जस अहई। सो तेहि भाँति रहे सुख लहई।।

*The Ocean said to Ram that 'O Ram! sky, air, fire, water and earth are inherently without consciousness. For the creation of the world, Maya, at the inspiration of Bhagwan created these elements. All scriptures say so. A person attains happiness by fulfilling those duties which are ordained for him.*

The Ocean further said, मरजादा पुनि तुम्हरी कीन्ही (That is, nature follows the rules made by You. I have followed these rules only. Then how far is this kind of anger justified? Only after it the Ocean tells Ram that, ढोल गँवार शूद्र पशु नारी। सकल ताड़ना के अधिकारी।।

Before discussing this *Chaupai*, we have to understand that the *Ramcharitmanas* was composed in Rajpur and Chitrkoot of Banda district. All the scholars agree that in the *Ramcharitmanas*, along with the Khadi dialect, Awadhi, Brij and Bundeli dialects have been freely used. It is natural that meaning of the *Chaupai* should be interpreted as per the circumstances and language, otherwise true meaning will be lost. It is evident that the Ocean was the ancestor of Ram. Ram, being a *Maryada Purushottam*, never violated any limit. Then, how could Ram have insulted the Ocean who is his ancestor?

If *Maryada Purushottam* Ram could not violate these rules, how could he who was always abided by dignity expect the Ocean to violate its dignity? Although, the Ocean had kept his self-respect aside, yet, it clearly told Ram that it has followed the rules and limits set by Ram himself for nature. Reminiscing about the rules of nature, the Ocean said in clear words that a person with power should do the welfare of the lesser or dependent persons of the society by keeping a special watch on them. In this *Chaupai*, the word 'ताड़ना' (thrashing) is of Bundeli language, which literally means to take special care of. If a villager, talking in Bundeli language, asks another person to take care of someone or something, he will say 'भैया इसे ताड़े रइयो' (Brother, take care of this). In this *Chaupai*, the word *'Tadana'* has to be understood in the cultural context of the Bundeli dialect. When an elder says something to a child or a young man of the next generation, he says it by the way of explanation. If we consider the word 'chastisement' of Bundeli language, the meaning of this *Chaupai* will be according to the context and according to the inherent motive of public welfare. For example, everyone knows that drum will burst if its strings are kept too tight during rainy season. Loose strings will have to be tightened so that proper sounds come from the drum. Similarly, a person with poor intelligence, if not looked after properly may cause harm either to himself or somebody else. He, being endowed with little intelligence does not have the ability to make decisions on his own properly. Hence he needs special care. In *Bhartiy* literature, *Shudr* is considered as an employee or a servant. The *Dharmik* scriptures have entrusted the entire concern and welfare of an employee or the servant to the employer or master. A non-caring attitude of master or an employer cannot bring happiness in the life of an employee or servant. Therefore, such persons require special attention. Similarly, an animal tied to a peg survives on the grace of the owner. Animal will not be able to survive if it is not taken care of by its owner. Hence, it needs special care. The physical structure of a 'female' is biologically different from that of a male. Females have less physical strength than males. Due to the physique she is considered to be the worthy of special protection in the society.

This speech of the Ocean to Ram is actually a reference to the laws of creation. Unfortunately, Hindu-phobics use the misinterpretation of this *Chaupai* as a weapon to malign the spirit of great public welfare present in the *Ramcharitmanas*. They conspire to affect public opinion against Ram and spread unrestrained propaganda.

Astu! many *Bhartiy* texts have been interpreted by individuals for the fulfillment of their wishes. More often than not, intellectual level of the interpreter is visible in the interpretation. With a change in position of *Comma* or any other grammatical sign the meaning changes completely.

| | |
|---|---|
| रोको, मत जाने दो। | *Stop, not allowed to go* |
| या | *or* |
| रोको मत, जाने दो। | *Stop not, allowed to go* |

Same words have been used in both the sentences but with a change in the placement of *Comma* the meaning of the sentences become contradictory. Apart from this, in Hindi language the same word has many meanings, it is called *'Shalva Alankar'*. The desired meaning can be derived by only shifting a word or symbol. Due to this reason, the present contentious *Chaupai* got entangled into controversies, otherwise the correct and true meaning would be clear, by understanding the text in its totality. I am confident that after going through my thoughtful deliberations saints and learned persons, in social interest, will cooperate in destroying the planned and vicious warp and weft of conspiracies against Hindu society.

The episode of beheading of Shambuk is also targeted. Because of this incident, Ram is criticized and the *Ramayan* is crucified. In fact, as I have said earlier, some conspirators with perverted mentality, tampered with the scriptures and added some conjectures so that Hindu theology and Hindu theosophical thought can be proved to be anti-*Dalit*, and thereby a large section of Hindu society gets cut off from its very roots. As a result of these conspiracies, the Hindu society has become weak. In any of the scriptures there is no prohibition on performance of austerities or *Yagy* by a *Shudr*. The *Vedic* literature from the *Purans*, the *Upnishads*, the *Bhagwad Gita*, the *Ramayan* etc., do not endorse to the *Varn System* based on birth. A holistic reading of the *Ramayan* and other texts makes it impossible to believe that Shambuk was punished by Ram as he was performing austerities in spite of being a *Shudr* by *Varn*.

The *Manusmriti*, considered to be the most controversial text, has given the right to all to perform austerities even without reciting of *Mantrs*.

धर्मेप्सवस्तु धर्मज्ञाः सतां वृत्तिमनुष्ठिताः।
मन्त्रवर्जं न तुष्यन्ति प्रशंसां प्राप्नुवन्ति च।। 10,120

*In the above Shlok, it has been said that Shudrs are praised for performing religious rituals even without reciting Mantrs. It means that the Manusmriti not only gives the right to the Shudrs to perform religious rituals and austerities, but also inspires them to perform religious rituals. (10,120)*

The intent of the *Manusmriti* can be inferred that a *Shudr* by *Varn*, if lagging behind in society due to lack of knowledge can also assume another *Varn*; he can also be a *Dwij*. A wrongly pronounced *Mantr*, instead of benefits would inflict harm. Inspite of this the rituals of *Yagy* could be performed by *Shudrs* because intention is equally important. What could be better than this? If, for argument sake, we agree that *Shudrs* did not have the right to recite *Mantr* then why are the *Mantrs* composed by all *Shudr* clans find place in the *Veds* and are an integral part of the *Veds*? Why are there several such sayings in the scriptures as 'आत्मवत्–सर्वभूतेषु (Think of all beings as your own self) सर्वभूतहितेरताः (for the benefits of all beings) ईशावास्यमिदं सर्वम्' (everything in the entire universe is pervaded by God)?

Discussions of the *Veds*, *Purans*, *Upnishads*, *Gita* etc. make it crystal clear that the *Varn System* was not based on the birth, but was based on attributes and *Karm*. It proves that incident of beheading of Shambuk in the Valmiki *Ramayan* is interpolated. *Shudr* Shambuk could not have been beheaded by Ram, who ate the left over Jujube berries of Shabari. Not only this, Ram could not have discussed नवधा भक्ति (nine types of devotions) with Shabari. Ram of the Valmiki *Ramayan* even asked Shabari : कच्चित्ते निर्जिता विघ्नाः कच्चित्ते वर्धते तपः।। रामायण 74,8 (Have all your obstacles perished? Are your austerities increasing) Shabri replied: अद्य प्राप्ता तपः सिद्धिस्तव सन्दर्शनान्मया। रामायण

74,11 (Today, by the grace of your *Darshan*, my *Tapshchrya* / austerities have been realized).

The question is, when Ram discusses *Tapshchrya* with *Shudr* Shabari, and also encourages her for it, then why would he kill *Shudr* Shambuk for doing *Tapshchrya*?

Whether it is Valmiki's or that of Tulsi's Ram, Ram is very kind, dutiful and *Maryada Purushottam*. He considers himself blessed by eating the jujube berries of Shabari. He embraces Nishadraj and addresses him as his dear brother like Bharat. Ram lives happily among the underprivileged and the oppressed for 14 years and treats them as his brothers. Will such a Ram kill anyone without any due reason? No! Absolutely not! An other episode of in the *Valmiki Ramayan* proves that it is incongruous. Shravan Kumar was the child of *Vaishy* father and *Shudr* mother. In the *Valmiki Ramayan* itself, he was addressed as an ascetic. Not only this, his parents have also been called ascetics. Shravan was filling his pot with water from a river and got shot down with an *Shabad-bhedi-baan* and he shouted : कथमस्मद्विधे शस्त्रं निपतेच्च तपस्विनि (how was an ascetic like me attacked?)

In this context he further said :

ऋषेर्हि न्यस्तदण्डस्य वने वन्येन जीवतः।
कथं नु शस्त्रेण बधो मद्विधस्य विधीयते।।
जटाभारधरस्यैव बल्कलाजिनवाससः।।

*Leaving everything behind, I was living like a Rishi, eating tubers in the forest. I have matted hair, wear Valkal and deer skin, then why am I being killed with a weapon?*

It is clear from the above statement that there was no restriction for *Shudrs* to have matted hair, wear *Valkal* and deerskin etc. In this context, Yuvraj Dashrath addressed Shravan as a *Rishi*. The following two *Shloks* of the *Valmiki Ramayan* should be duly appreciated :

कस्य वा पररात्रेऽहं श्रोष्यामि हृदयङ्गमम्।
अधीयानस्य मधुरं शास्त्रं वान्यद् विशेषतः।।
को मां सन्ध्यामुपास्यैव स्नात्वा हुतहुताशनः।
श्लाघयिष्यत्युपासीनः पुत्रशोकभयार्दितम्।। 64,32,33

*Shravan Kumar's father says in whose melodious voice during night, will I hear scriptures? From whose mouth will I hear a fascinating scripture-discussion? Who will now help me in taking bath, Sandhya Upasana and Agnihotr. Who will comfort, sitting beside this old man, suffering from the grief of son's death? (64,32,33)*

Astu! it is clear that Shravan Kumar used to study scriptures and used to discuss these with his parents. He also used to take bath, do *Sandhya Upasana* and *Hawan*. Traditionalists do not consider *Varnsankar* to be pure, whereas in the same episode of the *Ramayan*, the father of Shravan Kumar says :

नहि तस्मिन् कुले जातो गच्छत्यकुशलां गतिम्।
स तु यास्यति येन त्वं निहतो मम बान्धवः।। 64,45

*No man born in the clan of ascetics like us can attain misfortune. The disaster will dawn on you who killed. (64,45)*

In the light of this incident, when we look at the incident of Shambuk's beheading, we find no similarity between the two incidents. When *Ramayan* gives all rights to *Shudr* Shravan Kumar, why contradictory views appear in the case of Shambuk? It is also clear that in the case of killing of Shambuk, a *Shudr*, subsequent interpolation have made it a case of *Shudr* killing. The summary of the disputed Shambuk's beheading case described in *Uttarkand* is as follows :

One day, a *Brahman*, aged about five thousand years, with the dead body of his child who had not attained puberty knocked at the royal gate. There is not an iota of truth in this story. Such a terrible incident as of young death was never seen or heard in the kingdom of Shri Ram. The question is, how many years did Ram's reign last? A five thousand years old *Brahman*! Not only this, directly attacking Ram, the old man says that due to some of Ram's great misdeeds, children living in his kingdom are dying. Accusing Ram, he said that the kingdom of the Ikshvaku dynasty of saintly kings has become an orphan in Ram's kingdom. The deaths of children have become a lodestar. In a decisive tone he says that his son died due to some sin committed by Ram. On hearing the accusations by a *Dwij*, Ram called an assembly of his brothers, the ministers, *Vashishth* ji, Vayudev and Narad etc. Ram described the incident and asked for the reasons of that child's death. Before this august assembly Narad says "In the golden age, only *Brahmans* had the right to austerity. Non-*Brahman* had no right. In *Tretayug*, when *Brahmans* and *Kshtriys* became equally powerful, then only *Brahmans* and *Kshtriys* have the right to do *Tapshchrya* and rest of the *Varns* do the work of service." Then the veracity of the above statements made by Narad is itself very doubtful as in the *Ramayan* itself *Shudr* Shravan Kumar is referred to as an ascetic.

Narad, in the end says that some evil minded and perverse are surely doing *Tapshchrya*, that is why the child has died. He also tells Ram that the king who follows the *Rajdharm* gets one-sixth of the portion of good deeds of his people. The main *Dharm* of the king is the protection of the people of his kingdom. According to the incident, Ram, after keeping the Brahman's dead son in oil, searches for such a person. Not finding anyone in north, east or west, Ram, riding on his *Pushpak Viman* searches in southern direction. On the bank of a huge lake situated in the north part of Shaival mountain he sees an ascetic immersed in *Tapshchrya*, hanging upside down. Praising his *Tapshchrya*, Ram inquires about the purpose of his *Tapshchrya*. Shambuk says that he is a *Shudr* and he wants to go to heaven in his body, therefore he is doing such a fierce *Tapshchrya*. He also says that he is doing *Tapshchrya* to conquer *Devlok*. Ram takes out the sword and beheads Shambuk.

Even if this incident is discussed as it is, Shambuk's *Tapshchrya* to go to *Devlok* in the body can be considered as the reason for Ram's action. There is another incident of *Vashishth* cursing King Satyavrat, the father of truthful Harish Chandra, who wanted to go to heaven in the body. As a result he had to become a *Chandal*. *Muni* Vishwamitr, with the power of his *Tapshchrya* wanted to send king Satyavrat in his body to heaven. Instead of reaching heaven king Satyavrat got stuck in between and he had to hang upside down. Astu! Shambuk's doing *Tapshchrya*, as a *Shudr*, is not a sin, but his excessive ambition to conquer *Devlok* and to go to heaven in body can be the reason for his beheading. It should also be understood that whenever someone used to do severe *Tapshchrya*, Indra-thought, that, that person wanted to grab his throne. Out of fear, Devraj Indra used to conspire against such persons doing *Tapshchrya*. In this story also, Narad's presence and linking the death of Brahman's son with Shambuk's *Tapshchrya*, and after Shambuk's beheading, Devraj Indr giving proof of the existence of Brahman's son, may be a part of Indra's conspiracy. Otherwise, an ideal king like Ram could never have done such an inhuman act. In the episode of Shambuk's beheading, *Varn* or caste has been propagated with the intent to destroy the basic spirit of the *Ramayan*. The basis of *Varn* or caste for beheading of Shambuk is not in sync with the ideals upheld by Ram.

Tulsi did not consider this incident to be true. That is why Shambuk's beheading is not discussed in the *Ramcharitmanas*. Bhavabhuti has discussed the slaughter of Shambuk in the Uttar Ramcharit. This incident is also described in the Anand *Ramayan*. But the interpolated incident of cruelty as described in *Uttarkand*, the Valmiki *Ramayan* is not to be seen anywhere else. Bhavabhuti's Shambuk, after beheading becomes a divine person. While praising Ram, he expresses his gratitude towards Ram and says :

शम्बूक एष शिरसा चरणौ नतस्ते, सत्सङ्गजानि निधनान्यपि तारयन्ति।।

*This Shambuk bows down at your feet. Death received from satsang lets one across the ocean.*

Bhagwan Shri Ram conveys :

तद्नुभूतयतामुग्रस्य तपसः फलम्।।

*May you get the fruits of your hard Tapshchrya.*

While addressing Shambuk, Shri Ram says :

यत्रानन्दाष्च मोदाश्च यत्र पुण्याश्च सम्पदः।
वैराजा नाम ते लोकस्तैजसा सन्तु ते शिवाः।।

*Where there is the pleasure of enjoying spiritual, transcendental objects and sacred wealth is present, may you get that luminous and auspicious Brahmlok.*

In the end, Sri Ram bids farewell to Shambuk with bright eyes :

भद्र! शिवास्ते पन्थानों देवयानाः। प्रलीयस्व पुण्येभ्यो लोकेभ्यः।।

*O Gentleman! May the path called Devyan be beneficial for you. Now you merge to attain the virtuous worlds.*

A discussion of the *Devyan* Path (transcendental pathways) seems necessary here. The *Gita* and the *Chhandogy*

*Upanishad* describe two paths after death. These two paths are the *Devyan* and *Pitriyan*. Following the path of self-knowledge and *Tapshchrya*, a self-realized devotee with selfless devotion, goes to *Brahmlok*, the abode of Gods through transcendental pathways and does not come back. *Moksh*, the ultimate goal of life is attained. Those who, with some purpose perform *Yagy* and other sacrifices, go to *Chandrlok*, abode of the moon through *Pitriyan*. When the merits of such persons get exhausted, they have to return to the mortal world. It is clear from this that Shri Ram has made it clear that there is nothing wrong in Shambuk's *Tapshchrya* by telling Shambuk to go through the *Devyan* route. Because of his purity and selfless spirit he was destined to attains *Moksh*, the ultimate goal of life.

The story of Shambuk is found in the Kedar section of the *Skandh Puran*. In Chapter 106, Shambuk, a resident of Vijaypur is said to have belonged to *Shudr* clan in *Tretayug*. He lived in the *Ashram* of a *Brahman*. He received great respect in the Brahman's *Ashram*, who was constantly engaged in auspicious works and service of pilgrims. Once, this Shambuk, by a bullock cart reached Gangadwar of Haridwar. During his travel he met many *Rishis*, kings and wealthy people. By living in a virtuous place and in the company of *Rishis*, he became sinless and learned. Subsequently, he died in the company of Bhagwan Shri Ram and attained *Moksh*. In this episode, described in the *Skandh Puran*, Shambuk is not killed by Ram, but gets *Moksh* in his holy company. In this story the religious conduct of Shambuk has been praised. He is said to be completely innocent and learned. In the interpolated part of the Valmiki *Ramayan*, Shambuk was doing *Tapshchrya* in the south of Ayodhya. According to Bhavabhuti, he was engaged in *Tapshchrya* in *Dandkarany*. In the *Skandh Puran*, he attains *Moksh* in the company of Shri Ram in Haridwar, which is in north of Ayodhya. Thus, the author of *Skandh Puran* also did not accept the interpolated version of this story in Valmiki *Ramayan*.

Description about his family and the beheading of Shambuk is also available in the Jain Ramayans 'पउमचरिउ' (Paumchariu) and 'स्वयम्भूदेव' (Swyambhudev) composed by Vimal Suri.

According to them, Shambuk was the son of Ravan's sister, Chandranakha and Vidyadhar prince Khardushan. After doing austerity for 12 years, he had obtained a sword named Suryahas. The moon also used to lose its effulgence because of its edge had the vision of time. Coincidentally at the same time Ram, Sita and Lakshman reached there during their exile. When Lakshman saw that wonderful sword, he took it in his hand and hit the clump of bamboos, behind which Shambuk was standing. Lakshman's intention was to examine the amazing power of that sword. Along with a clump of bamboos, Shambuk's head was also cut off by Lakshman. Chandranakha, after receiving the information of beheading of his son, went to Lakshman to take revenge. She was bewitched by the beauty of two princes of Ayodhya and requested for their courtship. On being refused by both the princes, she started instigating both her elder brother Ravan and husband Khardushan to wage war against Ram and Lakshman. She further said that if they did not end their life, then she would enter alive in Agni.

जे घइउ पुत्तु महु–त्रणउ। खर–णन्दणु रावण भायणउ।।
तहों जीविउ जइण अञञु हरमि। तो हुयवह–पुञञो पीसरमि।।

According to statements in the Jain *Ramayan*, Shambuk, the nephew of Ravan was a great *Pundit*, the knower of the entire *Veds*. He was not a *Shudr*. Ravan was born in a *Brahman* family. He was the grandson of *Rishi Pulastya*, so how could his sister's son be a *Shudr*? It is also clear from this that being a *Shudr*, Shambuk was not killed by Ram for doing *Tapshchrya*, but by Lakshman accidentally, without intention or deceit. Vimal Suri has given the time of the creation of his book, as 530 years after Mahavir Swami's Nirvan i.e. 60 BCE. Thus, the Jain story is quite old. So it cannot be ignored in the case of the beheading of Shambuk.

Astu! it is clear from this analysis that the story of the beheading of Shambuk, described in the Valmiki *Ramayan* is interpolated and is contrary to the ideals. Since Shambuk had not done any work against scripture by doing *Tapshchrya*, where does the question of killing him arise? This story does not match even with the holy character of Ram, who embraced the boatman; ate the jujube berries of Shabri; lived for 14 years in the midst of the most oppressed people of the society; who renounced Sita just because a small citizen commented against her, who sentenced his brother Lakshman to death for violating the king's order, who established *Ramrajy* for the benefit of all. How could such a Ram kill the innocent Shambuk engaged in austerity?

Through an analysis of other few episodes in *Valmiki Ramayan*, we can understand *Ramayan's* original approach towards *Shudrs*. Sixth *Shlok* of the *Balkand* of *Ramayan* describes the fact that all people of Ayodhya were extremely prosperous, happy and content with the study of *Veds*. People of all *Varn* practiced *Dharm*. I am citing the following *Shloks* which shows the prosperity of Ayodhya.

नाकुण्डली नामुकुटी नास्त्रग्वी नाल्पभोगवान।
नामृष्टो न लिप्ताङ्गो नासुगधश्च विद्यते। बालकाण्ड 6,10

*That is, there is no one in Ayodhya who does not wear kundals in the ears, a crown on the head and garlands of flowers around neck. No one has any shortage of anything, there is no one who is not self-satisfied after taking a bath and whose body is not smeared with fragrant sandalwood. Everyone, like a flower, bloomed in the company of such persons. (Balkand 6,10)*

नानाहिताग्निर्नायज्वा न शूद्रो वा न तस्करः।
कश्चिदासीदयोध्यायां न चावृत्तो न सङ्करः।। बालकाण्ड 6,12

*There was no one who had not performed Agnihotr or Yagy. no one was a thief or smuggler, Shudr or even Varnasankar. People of all Varns performed Yagy etc. (Balkand 6,12)*

नाषडङ्गविदत्रस्ति नावृतो ना सहस्त्रदः।
न दीनः क्षिप्तचित्तो वा व्यथितो वापि कश्चन।। बालकाण्ड 6,15

*There was not a single one who had not studied the six branches of the Veds i.e. Kalp, Nirukt, Grammar, Astrology and Shloks, did not live by the rules, did not give thousands of thousands in charity, was sloppy, orphaned, mad, eccentric or unhappy. (Balkand 6,15)*

वर्णेष्वग्रयचतुर्थेषु देवतातिथिपूजकाः।
कृतज्ञाश्च वदान्याश्च शूरा विक्रमसंयुताः।। बालकाण्ड 6,17

*All the Brahmans, Kshtriys, Vaishys and Shudrs were always engaged in the service of the Deities and guests, were always gratuitous and expressed it to, were ready to donate everything for the good of others, were brave and capable of doing the greatest. (Balkand 6,17)*

From the above *Shloks*, we get a visions of the social system of the *Ramayan* period. All people of that time were mighty, generous, benevolent, full of human qualities. All were prosperous and no one had any restriction for leading a good life. Worship, *Yagy*, rituals were not forbidden to *Shudrs*, but they were well-versed in the *Veds*. :

सर्वे वर्णा यथा पूजां प्राप्नुवन्ति सुसत्कृताः।
न चावज्ञा प्रयोक्तव्या काम क्रोधावशादपि।। बालकाण्ड 13,14

*Vashishth calls the craftsmen and the administrators of the Yagy and directs that people of all Varn should be well respected and honored, and no one, even in the heat of anger or huff should be disrespected.*

Further in this case, *Vashishth* ji while directing Sumant ji says :

निमन्त्रयस्व नृपतीन पृथिव्यां ये च धार्मिकाः।
ब्राहमणान् क्षत्रियान् वैश्याञशूद्रांश्चैव सहस्त्रशः।। बालकाण्ड 13,20

*All the religious kings on the earth, may these be Brahmans, Kshtriys, Vaishys and thousands of Shudrs, should be invited to attend the Yagy. (Balkand 13,20)*

These *Shloks* are wonderful examples of inclusivity. There was no distinction between *Varns* based on birth. Ram befriended Nishad Guh and treated him like his brother.

Another example from the Valmiki *Ramayan* is also noteworthy. After Ram's exile, Bharat was called from his maternal grandparents' house. After his arrival in Ayodhya, Bharat became aware of the whole incidents. After his father's *Shradh*, an agitated and sad Bharat decided to bring Ram back to Ayodhya. For this purpose, he appealed to the people that everyone should come along and urge Shri Ram to return to Ayodhya. This is well explained in the following *Shloks* :

तथा समुत्थाय कुले–कुले ते राजन्यवैश्या वृषलाष्च विप्राः।
अयूयुजत्रूष्ट्ररथान खरांश्च नागान् हयांश्चैव कुलप्रसूतान्।।
अयो0 82,32

*From every household, Kshtriys, Vaishys, Brahmans and Shudrs got up and prepared their high breed race horses, elephants, camels, donkeys and plowed chariots to go along with Bharat. Everyone was ready to walk together. There was no discrimination of any kind.*

Another *Shlok* confirming the same fact says that :

इत्येवं कथयन्तस्ते सम्प्रहृष्टाः कथाः शुभाः।
परिष्व जानाश्चान्योन्यं ययुर्नागरिकास्तदा।। अयो० 83,10

*The citizens of Ayodhya were traveling, talking and hugging each other with great joy. (Ayo 83,10)*

People of all castes and genders, young and old, male and female, were joyously shouting slogans to bring back Shri Ram. They were glad that they would be able to bring back their own beloved Shri Ram. During the journey everyone was hugging each other. It is clear from this that there was no feeling of high, low or untouchability. *Brahmans*, *Kshtriys*, *Vaishys* and *Shudrs* were equally a part of this journey. The castes involved in this journey have been described in *Shloks* from 12 to 16 of 83rd *Sarg* of Ayodhyakand. Manishilpi; potter; weaver; weapon maker i.e. blacksmith; Mayuraks, who make umbrella and fan from peacock feathers; carpenter, who cuts wood with saw; jeweler, who pierces pearls and gems; ivory-makers; lime-makers; fragrance makers; gold-smiths, blanket-makers, *Kahars, Vaidyas, Dhupaks*, liquor sellers, washermen, tailors, scavengers, acrobats, seafarers and *Vedic Brahmans*. In this way people from 23 castes participated in this *Yatra*. Most of these castes are now included in the scheduled castes in the constitution of India, At that time these castes were resourceful and intelligent.

To understand the social harmony of the world of the *Ramayan*, it is necessary to look at the exile period of Ram. In the *Ramayan*, the castes which supported Ram the most are *Vanar* (Monkeys), *Riksh* (a type of bear) *Bhalu* (Bear), *Gidh* (Vultures), *Kirat*, *Kol* etc. In fact, these castes are found amongst tribals. *Nishad* and *Kewat* etc. too belong to the deprived classes. Father Kamil Bulke wrote on page 61 of his book, *the Ramkatha: Genesis and Development,* "The monkeys, *Riksh* and the demons of *Ramkatha* were the tribal non-Aryan races of Vindhya Pradesh and Central India." Difference of opinions is not found on this point. Although, in the *Valmiki Ramayan* these tribals are actually called *Vanar, Riksh* etc., yet at many places *Adi Kavya* shows that in the beginning they were all considered to be human beings. The *Vanars* of the *Ramayan* are intelligent like humans, speak human language, wear clothes, live in houses, lead family life and live under the rule of the king. It is clear from this that in the eyes of the poet, they were not a mere monkeys. They had their own culture and social system. In fact they belonged to the *Vanar, Riadi* tribe.

The famous scholar C.V. Vaidya believes that *Vanars* looked like monkeys that's why they were called monkeys.

In the conclusion, Father Kamil Bulke wrote that, "Keeping all these things in mind, it is clear that the tribals definitely have a relationship with the *Ramayan* and it seems more likely that the *Vanar, Riksh* and *Gidh* were *Gautriy* tribals."

According to Jain texts, each caste (tribes) had its own flag. These tribes, during their journey used to carry flags with some mark inscribed on it and these tribes were known by that mark. If there is a mark of the *Vanar* of the flag, then those people were called the *Vanars*. Be as it may, the people with whom Ram stayed were the people from lowest ladder of the society. These were the very people who helped Ram, to conquer Lanka and make the land free from demons and devils.

During Ram's search for his kidnapped wife, Sita, he saw the injured Jatayu, who had fought Ravan to rescue Sita. Fighting valiantly, Jatayu killed Ravan's charioteer, broke Ravan's bow and arrow. An infuriated Ravan cut off his wings with his sword and the incapacitated Jatayu fell on the earth. After giving details of Sita's kidnapping to Ram, Jatayu breathed his last. On the death of Jatayu, a saddened Ram said to Lakshman that:

राजा दशरथः श्रीमान् यथा सम महायशाः।
पूजनीयश्च मान्यश्च तथायं पतंगेश्वरः।। अरण्यकाण्ड 67, 26
एवमुक्त्वा चितां दीप्तामारोप्य पतगेश्वरम्।
ददाह रामो धर्मात्मा स्वबन्धुमिव दुःखितः।। अरण्यकाण्ड 68,31
शास्त्र दृश्टेन विधिना जलं गृधाय राघवौ।
स्नात्वा तौ गृध्रराजाय उदकं चक्रतुस्तदा।। अरण्यकाण्ड 68,36

*Pakshiraj Jatayu is as much honorable for me as King Dasharath was. He ordered Lakshman to bring dry wood so that fire could be ignited for Jatayu's cremation. Ram, like a son, performed the last rites of Pakshiraj Jatayu by lighting the wooden pyre. At the time of the last rites of Pakshiraj Jatayu, he wished that, 'O Mahabali Pakshiraja! You may also attain excellent abode which are attained by Yagyakarta, Agnihotri, men who do not show their back in battle field and who donate land. For Pakshiraj Jatayu, those very Mantrs, which were considered necessary to be recited for the attainment of heaven by ancestors were recited by Ram.*

Thereafter, Ram and Lakshman bathed in the Godavari river and offered *Jalanjli* (water offering) to Pakshiraj Jatayu according to the method prescribed by *Shastrs*.

From this episode one can easily see the greatness and goodness of Ram. Ram, cremated and performed all other last rites for Jatayu and offered *Tarpan*. According to *Shastr*, *Tarpan* can be performed only by son, brother-in-law and brothers. This dutiful act of Ram destroyed all the taboos of high and low.

Similarly, in the *Ramayan*, the importance of Sumantr, son of charioteer, as minister of King Dasharath is visible. There are many instances in the *Mahabharat* of repeated insults of Karan by addressing him as *Sut-Putr* (Son of charioteer) even though he was not so. He was only brought up by a charioteer. In the *Ramayan* it is not derogatory to be a *Sut-Putr*. *Sut-Putr* Sumantr, was very influential in Ayodhya's council of ministers and was respected too. This respect of *Sumantr* appears only in the 18th chapter of *Balkand* of the Valmiki *Ramayan*, where King Dasharath addresses Sumantr as a great minister. It is clear from this that the place of Sumantr is paramount in the Council of Ministers. He uninterruptedly moved in the palace of king Dasharath. How influential Sumantr was he is known from the conversation he had with Queen Kaikeyi and even could reprimand her. Consider the following *Shloks* in this regard:

यस्यास्तव पतिस्त्यक्तो राजा दशरथः स्वयम्।
भर्ता सर्वस्व जगतः स्थावरस्य चरस्य च।।
नह्यकार्यतमं किंचित्तव देवीह विद्यते।
पतिघ्नीं त्वामहं मन्ये कुलघ्नीमपि चान्ततः।।
यथावयोहिराज्यानि प्राप्नुवन्ति नृपक्षये।

इक्ष्वाकुकुलनाथेऽस्मिंस्तं लोपयितुमिच्छसि।।
राजा भवतु ते पुत्रो भरतः शास्तुमेदिनीम्।
वयं तत्र गमिष्यामों यत्र रामो गमिष्यति।।
नूनं सर्वे गमिष्यामो मार्ग रामनिषेवितम्।
त्यक्ता या बान्धवैः सर्वेब्राह्मणैः साधुभिः सदा।।
आभिजात्यं हि ते मन्ये यथामातुस्तथैव च।
नहि निम्बात स्रवेत् क्षौद्रं लोके निगदितं बचः।।
मातरं ते निरस्याशु विजहार कुबेरवत्।।

*He tells Queen Kaikeyi in clear words that When you have abandoned your mighty husband, then there is no such misdeed in this world which you cannot do. I consider you as husband's murderer and murderer of the clan. He also says that it has been a tradition of the Ikshvaku clan that after the death of the king, only the eldest son takes over the throne, but you have broken that tradition. You may make your son the king, but we will go where ever Ram will stay. All kinsman and scholars will leave your kingdom and go away. He also says that just like you, your mother was stubborn too. That is why, on the advice of a Rishi, your father had thrown your mother out of the house.*

Astu! It is evident from the above context that the only person who could reprimand the queen of Dasharath in such strong words must be of great importance in the eyes of the king and the subjects. His acceptance and respect in Dasharath's family was to such an extent that Ram while going into exile gave serious instructions to him for the welfare of King Dasharath, the state and the subjects. He said :

इक्ष्वाकूणां त्वया तुल्यं सुहृदं नोपलक्षये।
यथा दशरथो राजा मां न शोचेत तथा कुरु।। अयो० 52,22

*Sumantr ji, in my view, there is no one like you who is a well-wisher of the Ikshvaku clan. You do whatever you can so that King Dasharath does not grieve for me.*

Although, Sumantr urged Ram that he would also spend 14 years in exile with him, Ram did not allow him. In fact, Sri Ram had unwavering faith in Sumantr. That is why he sent him back and gave many massages for mothers, brothers and other family members.

मम प्रियार्थं राज्ञश्च सुमन्त्र त्वं पुरीं व्रज।
सन्दिष्टश्चापि यानर्थास्तांस्तान् ब्रूयास्तथा तथा।। अयो० 52,64

*You must return to Ayodhya to serve me and for the welfare of King Dasharath and narrate my messages.*

The immense respect for Sumantr is also expressed in this *Chaupai* of the *Ramcharitmanas*, in which Ram, seeing Sumantr, respected him like his father. *Chaupai* is like this : 'राम सुमंत्रहिं आवत देखा। आदर कीन्ह पिता सम लेखा।।' From these *Shloks*, it is clear that Sumantr was respected everywhere.

It can easily be understood from these episodes of the *Ramayan* that during the *Ramayan* period, every person of the society, irrespective of class or *Varn*, had equal respect and equal rights. There was no discrimination amongst people on the basis of *Varn*. In the light of these incidents, Ram's killing of Shambuk for performing *Tapshchrya* also seems to be a complete untruth.

Astu! After a deep study and from bird's eye view, it can be asserted that the *Ramayan* period is a true reflection of *Sanatan* Hindu society which was grand, sans taboos and was beyond all criticism.

# References from Jainism

The climax of spiritual practice and austerity in *Bhartiy* contemplations and philosophy can be seen in most of the Jain *Rishis*. Mahavir, addressed with the epithet of Bhagwan, was the last and 24th *Tirthankar* of the Jain sect. He himself was an embodiment of *Tapshchrya* and spiritual practices. Bhagwan Mahavir, the real incarnation of compassion and love, renounced the majestic splendor and adopted the path of meditation. Jainism is believed to have originated from Rishabh Dev, whose reference can be found in the *Rigved*, Bhagwad philosophy, *Purans* and *Upanishads*. There are 24 incarnations of Bhagwan Vishnu in *Sanatan* philosophy. Bhagwan Rishabh Dev, the 18th incarnation of Vishnu, is believed to have been born to Merudevi, wife of Nabhiraj. It is mentioned in the Bhagwad :

अष्टमें मेरु देव्यां तु नाभेजाति उरुक्रमः ।
दर्शयन् वर्त्म धीराणां सर्वाश्रमनमस्कृतम् ।। 1,3,13

As the 18th incarnation, the Bhagwan incarnated from the womb of Nabhiraj's wife Merudevi. Rishabh Dev showed the path of the *Ashrams*, worshiped by all the Paramhans. Rishabh Dev was followed by 23 more *Tirthankars*, of whom Bhagwan Mahavir is considered to be the 24th and the last. All the Jain *Tirthankars* have given sermons for the benefit of all beings, and not only for human beings. According to him 'मा हिंस्यात् सर्वभूतानि' (do no violence to any living being). All are equal in Bhagwan Mahavir's eyes. He neither talks of worshipping any *Brahman* nor does he talk of hating a *Chandal*. According to him, he, who gets victory over his vices by being impartial is worthy to be called a Jain.

Amongst all the *Tirthankars*, Bhagwan Mahavir is of utmost significance. On the *Chaitr Trayodashi* of *Shukl Paksh* of 586 B.C., Bhagwan Mahavir was born as the son of King Siddharth and Queen Trishala. Soon after Queen Trishala's getting pregnant, there was a qualitative increase in the glory of king Siddharth. That is why he was called Vardhaman. Many stories are prevalent about his supernatural feat. One of the stories is about Indr himself naming Bhagwan Mahavir as Vardhaman. The Digambars of Jain sect consider Bhagwan Mahavir to be a lifelong celibate, while according to the Shwetambar tradition, he was married to a girl named Yashoda, the daughter of Samant Samarveer. Under the Sal tree, on the tenth day of *Shukl Paksh* of *Vaishakh* month he attained knowledge of *Kaivaly*. This is the reason why he is called Kevlin. He was called Jin for having conquered the *Ashtkarms*. To propagate Jainism he continuously travelled for 42 years.

Bhagwan Mahavir preached social harmony through his principles and proclaimed the determination of the *Varn* on the basis of distinctive attributes and *Karm*. The *Uttradhyayan* has an important place in the exponent texts of Jainism. In its 25th study, it is clearly described that the basis of *Varn* is *Karm* and not birth. According to the context described in this study, two *Brahmans* named Jaighosh and Vijayghosh lived in Varanasi. One day Jaighosh saw a snake on the bank of the river being eaten on one side by a frog and on the other side by the Kurar bird. Due to intensified compassion in his mind, Jaighosh became a *Sanyasi*. Because of his vow for one month *Upwas* (fast) he stayed outside the city. On completion of the *Upwas*, he went to the city to seek *Paran* (to seek alms after *Upwas*). He saw Vijayghosh performing a *Yagy*. Jaighosh begged him for alms. On this Vijayghosh said that the sacrificial food is given only to those *Brahmans* who have knowledge of the *Veds* including Vedang and are *Brahmans*. After the rejection of the request, Jayaghosh described the characteristics of a *Brahman* and declared that :

न वि मुण्डिएण समणो न ओंकारेण बम्भणों।
न मुणी रण्णवासेणं कुसचीरेण न तावसो।।
कम्मुणा वंभणो होइ कम्मुणा होइ खत्तिओ।
बइसो कम्मुणा होइ सुद्दो हवइ कम्मुणा।।

उत्तराध्ययन 25–33

*That is, by shaving the head, hearing or chanting Oiu, one does not become a Brahman. Living in a deserted forest does not make one an ascetic. Just by wearing the clothes made from Kush one does not become a Tapasvi. Human beings are Brahmans by action, Kshtriys by action, Vaishys by action and Shudrs by action.*

Many stories in Jain literature clarify that no person is ignoble by birth. One becomes worthy of worship or hatred only by one's distinctive attributes and *Karm*. Accordingly, there is acceptance or rejection of an individual in the society. Many such incidents in the life of Bhagwan Mahavir exhibit his unflinching love for every creature. According to a legend, as a *Parivrajak* (Monk) Vardhaman was going somewhere, a *Chandal* named Harikeshbal came from the opposite direction. Seeing Harikeshbal's fearlessness and intensity, people shouted, "Stop him, lest he touches Vardhaman". Vardhaman stopped people from catching hold of Harikeshbal and asked them to let Harikeshbal come to him. As Harikeshbal bent down to touch Vardhaman's feet, Vardhaman hugged him and said, "Everyone's soul is one, there is no difference amongst human beings. This discrimination is a social hypocrisy." When Vardhaman became Mahavir, a *Tirthankar*, he initiated this *Chandal* Harikeshbal and made him a *Shraman* and gave him a place in his commune. This Harikeshbal *Chandal* became a person imbued with great qualities. Great reverence for Harikeshbal has been expressed in Jain scriptures. His qualities have been appreciated abundantly in these religious texts.

It is clear from these quotes and legends that Jainism completely refutes discrimination and is a proponent of a sensitive and deep philosophy and amity towards all creatures. Like other sects, Jainism is a sect of *Sanatan Dharm*. Unfortunately, some selfish and less knowledgeable people are engaged in separating Jainism from Hinduism. It is a part of malicious design to weaken Hindu *Dharm*.

The misconception that Jainism is considered to be a branch of Buddhism itself, is now cleared. In fact, the Jain sect existed much before the emergence of the Buddhist sect. This confusion arose as Jainism was less prevalent and Buddhism grew rapidly. From the time of Bhagwan Mahavir, the last and 24th *Tirthankar*, Jain sect expanded further. Some scholars consider the origin of Jainism from Bhagwan Mahavir while some scholars consider Bhagwan Parshvanath, born 250 years before Christ, to be founder of Jainism. However, according to some, Jainism had the emerged from Rishabh Dev. According to the scriptures of *Sanatan Dharm*, Bhagwan Rishabh Dev is the first *Tirthankar* and founder of the Jain sect. Many scholars accept this view.

# References from Buddhism

Buddhism emerged following severe *Tapshchrya* by Prince Siddharth of Kapilvastu, resulting in his enlightenment under the Bodhi tree on the banks of the Niranjan River. It is said that during the ceremony organized on the birth anniversary of King Shuddhodhan, astrologers, seeing the baby Siddharth predicted that this child would be either the hero of the world, the remover of sorrows, the enlightened sage or would be known as *Chakrvati* (emperor). *Sanatan Dharm* believes that the incarnation of Bhagwan is for the salvation of the oppressed. Prince Siddharth, the embodiment of the ocean of compassion and benevolence, was born for the protection of gentle and poor people and that is why he abandoned the majestic retinue and finery. He severed his ties with his beautiful wife and newborn son and followed the path of severe *Tapshchrya* in the pursuit of Truth. After attaining supreme knowledge and *Nirvan*, he, for 45 years continued to spread the light of the knowledge. He, who could not be deterred by bandit chief Angulimal, who had brutally killed 999 people, was extremely disturbed by the foul-smelling physique of a sick beggar. After washing and cleaning the beggar with his own hands, the prince became a monk. The awakening speech of Bhagwan Buddh on this occasion is important : "Arrival of Tathagata (awakened Buddha) in this world is meant to befriend the poor, to help the vulnerable, to heal the diseased from sufferings, to enlighten the ignorant and the confused; to protect the rights of the orphans and the old, by performing such actions the *Tathagata* sets an example for others to follow. By doing so they themselves become an example for others."

For his philosophy, based on extreme compassion and love he is respected everywhere as the savior of suffering humanity across castes and *Varns*. He told to king Bimbsaar :

पदे तु यस्मित्र न भीर्न रुड्. न जन्म नैवोपरमो न चाधयः।
तमेव मन्ये पुरुषार्थमुत्तमं न विद्यतेपत्र पुनः पुनः क्रिया।।

*That is, the one who has neither old age, nor fear, nor disease, neither birth or death, nor illness, that is what I consider to be the best Purusharth, in which there is no action of frequent movement.*

Many scholars like Edwin Arnold, Pandurang Vaman, Kale, philosopher Nagarjun etc. have described Bhagwan Buddh and the principles propounded by him as the basic universal *Mantr* of liberation.

Bhagwan Buddh propounded and propagated the principle of social equality in life. He wandered around and chanted that there is no *Brahman* or *Shudr* by birth. The Dhammapad contains a collection of 423 *Gathas* (Stories) narrated by Buddh from time to time, which are fundamental for understanding the tenets of Buddhism. In fact, he has given more importance to *Karm* and Gyan than to the birth in *Varn System*. These thoughts are stored in the *Tripitk* texts in the Pali language.

There is a context in *Vaseth Sut* i.e. when Bhagwan Buddh was residing in Ichha Nangal, there was a dispute between two *Brahman Rishis, Vashishth* and Bharadwaj on whether a *Brahman* is by birth or by attributes and *Karm*. When the dispute between the two got intense, both went to Bhagwan Buddh for the solution of their doubts and said :

"O Gautam!, there is a dispute between the two of us whether birth is the basis of *Varn* or *Karm*?" The answer given by Bhagwan Buddh regarding the *Varn System* is in line with the *Sanatan Shastrs* in which the *Varn System* is not based on birth but on the inherent individual attributes and *Karm*. He said - "There is a difference in the species of animals like insects, moths, ants, small and big quadrupeds, species of snakes, fish and birds, on the basis of these differences, they have different identities. There a no such distinction based on species amongst humans."

He further said "Such a distinguishing mark is neither in the hair, nor in the head, nor in the ear, nor in the eyes of human beings."

Neither in hair, head, ear eyes, mouth, nostrils, lips nor in eyebrows. Neither in throat, shoulders, back, stomach, hips, place of worship, nor in sex, neither in hands, feets, nor in nails.

*Neither in thighs, chest, Varn, nor in Swarn. There is no distinguishing mark in human body nor in human spices. (17) The difference in human beings is in Consciousness. (18) O Vashishth! Those who earn their livelihood from cattle, know them to be Agriculturist not Brahman. (19) O Vashishth! Those who earn their livelihood from craft, know them to be Craftsman not Brahman. (20) O Vashishth! Those who earn their livelihood from trade, know them to be Vaishy not Brahman. (21) O Vashishth! Those who earn their livelihood from remittance, know them to be Servants not Brahman. (22) O Vashishth! Those who earn their live on donations, know them to be Thieves not Brahman. (24) O Vashishth! Those who earn their livelihood from Weapons, know them to be Warriors not Brahman. (25) O Vashishth! Those who earn involved in priesthood, know them to be Clergy not Brahman. (26) O Vashishth! Those who live by village/kingdom, know them to be Kings not Brahman.*

After this, there is a discourse in the *Dhammapad* about who is a *Brahman*. The summary is that no one is *Brahman* or belongs to any other *Varn* by birth.

In *Sanatan Dharm* Bhagwan Buddh is praised as an incarnation of Bhagwan Vishnu. A few texts hold Bhagwan Buddh as the ninth incarnation of Bhagwan Vishnu while few other texts refer to him as the tenth incarnation. It may be disputed weather he was ninth or tenth incarnation but it is indisputable that he is an incarnation. What is meant here is that in *Sanatan* literature or Hindu *Dharm*, Bhagwan Buddh is considered to be an incarnation of Vishnu. In *Sanatan Dharm*, not merely Buddh has been praised as Bhagwan; his reference in rituals is not only essential but is also considered indispensable. This *Sankalp Mantr*, which is read by the priest on every auspicious occasion, is a clear proof of this fact :

ऊँ अद्य ब्रह्मणोपि द्वितीयपरार्धे श्रीश्वेतवाराहकल्पे वैवस्वतमन्वन्तरे
अष्टाविशंतितमे कलियुगे कलि–प्रथचरणे बौद्धावतारे भू लोके जम्बूद्वीप
भरतखण्डे आर्यावर्त देशान्तर्गते पुण्यक्षेत्रे

This resolution is compulsorily recited aloud by the priest in any worship performed by the Hindu society throughout Bharat. The host of the ritual is immediately consecrated. From this it is quite clear that Bhagwan Buddh is an inseparable part of *Bhartiy* thinking, philosophy and spirituality.

Jagadguru Adi Shankrachary, Swami Vivekanand, etc. and all modern *Sanatan Rishis* have also considered Bhagwan Buddh to be the soul of *Bhartiy* life and an inseparable part of Hindu philosophy.

Readers must appreciate Swami Vivekananda's sentimental expression regarding Bhagwan Buddh :

"Bhagwan Buddh is my presiding deity, my Bhagwan. We worship him as an incarnation of Bhagwan Vishnu. Such a fearless propagator of morality has never arisen in the world. Bhagwan Krishan, the best among *Karm* Yogis, is born again as a disciple to culminate His teachings into action. The same voice which had guided Arjun in the *Shrimad Bhagwad Gita* was heard again. As a living example of the teachings of the *Gita*, the preacher of the *Gita*, Bhagwan Shri Krishn came again in this mortal world in the form of Shakaymuni, Abandoning the throne, he started living with the unhappy, poor, fallen, beggars. Like Ram, Prince Gautam also hugged *Chandals*."

Guru Gaudpadacharya, the grandfather of Adi Shankrachary, has also praised Bhagwan Buddh. Eulogizing he hailed Bhagwan Buddh as the ace human being. Later Adi Shankrachary considered Bhagwan Buddh as the ninth incarnation of Vishnu and lauded him. Adi Shankrachary's hymn are stored in 'विष्णुपादादि के शान्त वर्णन' (Quiet description of Vishnu Padadi) published by Sri Venkateswar Press Mumbai. *Shlok* no. 50 describes ten incarnations out of which Bhagwan Buddh is revered as the ninth incarnation :

मत्स्यः कूर्मो वराहो नरहरिणपतिर्वामनो जामदग्नयः।
ककुत्स्थः कंसधानी मनसिजविजयी यश्च कल्की भविष्यन्।।
विष्णोरंशावतारा भुवनिहितकरा धर्मसंस्थापनार्थाः।
पायासुर्मां त एते गुरुतरकरुणाभारखित्राशया ये।।

*Fish, Tortoise, Boar, Nirsingh, Dwarf, Parshuram, Ram, Krishn, Buddh and Kalki who will incarnate in the future, are incarnations of Vishnu for the welfare of the Varns and the establishment of Dharm. May all these be full of compassion and protect me.*

The *Puranic* scriptures not only acknowledge Bhagwan Buddh but on the occasion of *Pind Daan* and *Shradh* at Gaya for the salvation of ancestors also make mandatory the ritual of worshipping the *Peepal* tree under which Bhagwan Buddh had attained enlightenment. The *Mantr* to be recited at the time of *Pind Daan* in Gaya is as follows :

नमस्ते श्वत्थराजाय ब्रह्माविष्णुशिवात्मने।
बोधिद्रुमाय कर्तृणां पित्णां तारणाय च।।
ये स्यत्कुले मातृवंशे बान्धवा दुर्गतिं गताः।
त्वदर्शनाच्च, स्पर्शाच्च स्वर्गतिं यान्तु शाश्वतीम्।।
ऋणत्रयं मया दतं गयामागत्य बृक्षराट।
त्वत्प्रसादान् महापापाद् विमुक्ताऽहं भवार्णवात्।।

वायु पुराण 111,28&30

*I bow to Ashwatthraja, who is in the form of Brahma, Vishnu and Mahesh. These are the Ashwatthraja Bodhi trees. I salute them for the salvation of my forefathers. In our family or in the family of our mother, the brothers who have met catastrophe, will attain salvation by your darshan and touch. O King of Trees! After coming to Gaya, I have repaid all my three debts. By your grace, I have been freed from the great ocean of sins.*

Apart from this, in many scriptures like the *Matsy Puran*, *Agni Puran*, *Bhagwad Puran*, *Brahm Puran*, *Bhavishy Puran* etc., Bhagwan Buddh has been praised as an incarnation of Vishnu.

It is noteworthy that in the scriptures of *Sanatan Dharm*, Bhagwan Buddh has been considered as an incarnation which simply means that Buddhism is an off shoot of Hinduism or *Sanatan Dharm*. In the present times, in antagonism of Hindu thought, a few take Buddhism either to be different from Hinduism or not as one of its sects. It must be said that they are engaged with full force in proving that the Buddhism is an opponent of Hindu *Dharm*. The followers of different ideologies within Hinduism have also started seeing Buddhist sect as anti-Hindu *Dharm*, whereas in Hindu *Dharm* or *Sanatan* philosophy it is believed that to protect the gentle people and to re-establish *Dharm* Bhagwan will incarnate in every Yug (age) whenever there is any loss to *Dharm* or unrighteousness increases, or people without conduct dance in fury, at the time Bhagwan will incarnate. *Dharm* had been perverted. Animal sacrifice was at its peak. During Buddh's time, the proclamation of non-violence and opposition to evil practices was the human duty as the Yug*dharm* and Bhagwan Buddh fulfilled that duty. He confronted all those who blindly followed the stereotypes. Ultimately the truth was victorious and Buddh became Bhagwan. The Hindu scriptures accepted him as an incarnation and he has been worshiped accordingly as Bhagwan. The need of the day is that the sublime principles of Bhagwan Buddh should be expounded in the light of the Hindu scriptures. Whoever considers *Sanatan Dharm* and culture to be indispensable for world humanity, they will have to take steps in this regard with a well-planned strategy and determined mind.

Buddhist philosophy also appears to stand out in support of the theme of the present book. According to the Buddhist philosophy propounded by Bhagwan Buddh, as an incarnation of Bhagwan Vishnu, there is no place for thinking against the *Varn System* as based on attribute and *Karm.*

Astu! according to the Buddhist philosophy, the *Varn System* is not based on birth, but it is based on individual's attributes and *Karm* as has been propagated in the *Sanatan Shastrs* too.

# References from Sikhism

Guru Nanak Dev is considered to be the precursor of the Sikh sect. The emergence of Guru Nanak Dev took place at a time when foreign invaders with their nefarious designs were trampling down the pride of the land of Bharat. Their aim was not only to rule and plunder but they also wanted to completely destroy and corrupt the Hindu *Dharm* and convert everyone to Islam. Anyone who resisted was unapologetically and inhumanly beheaded. By enslavement of girls, women and children; by shaving their heads; enchaining and dragging them like animals, humanity was being targeted. Hindu society had also gone astray. *Bhartiy* society had assimilated several malpractices. The Muslims were also engaged in killing, looting and committing adultery targeting Hindu girls and women and thereby crossing all limits of humanism.

In fact, all attacks of aliens on Bharat were from Jammu and Kashmir side, that is, from the north. Therefore, the first area to be trampled down was Punjab after Jammu and Kashmir. The people of these provinces had to suffer the most.

The Afghans and Turks established their kingdoms and various Muslim rulers and their representatives ruled on the basis of military power. They committed countless inhuman atrocities, imposed Jazya tax and other heavy taxes on non-Muslims to such an extent that *Bhartiys* got outraged. Their aim was to make everyone Muslim. The recruitment of Hindus in jobs was stopped. Those who were appointed were removed, orders were given to demolish the temples. They were converted into *Maqtabs* and *Sarais*. New mosques were built. Hindus were prohibited from applying Tilak. The construction of new temples was not only banned, but idol worship was also declared to be a crime. Hindus could not a sit on horse or in palanquin. They could only ride on an ass. The blasphemy law was strictly enforced. There was death penalty for saying anything against Islam. Sikandar Lodi sentenced a *Brahman* named Bodhan to death only because he had equated Islam and Hinduism. At a time when the morale of the Hindu society was completely depleted and the culture was dying, Guru Nanak, as a God incarnate, appeared on the horizon as saviour to protect *Bhartiy* society.

Guru Nanak was born in a Khatri family on 15 April 1468 in Talwandi, 65 km from Lahore. After leading a family life for 27 years, he attained self-realization and knowledge. Being of reflective nature, he lived in the company of both Hindu and Muslim seers. On the one hand, he was pained by the atrocities being committed by the invaders and on the other hand he sought to end the sectarianism through monotheism. By creating social awareness against both the evils, he laid the foundation of such a philosophy as would reflect the Hindu philosophy and would also block the flow of conversion. He laid the foundation of a new casteless and classless society through his philosophy. Like the other great men born in Hinduism, he made *Vedic* philosophy the basis of his thinking. His aim was to end evils, caste-discrimination and communal hysteria prevailing in the Hindu society and establish social harmony. He was able to fulfil his aim. The long line of Guru *Parmpara* made every effort to achieve this objective of Guru Nanak Dev. History is a witness to innumerable sacrifices made by the Sikh gurus for this. Some conspirators who were dreaming of eliminating the Hindu society wanted to project Sikhism as a separate sect, like Jainism and Buddhism. In fact, if all these sects are separated from Hinduism, there will be no existence of Hindu *Dharm* and this was the main goal of the anti-Hindu conspirators. While Guru Nanak Dev made a scathing attack on the evils prevailing in the society, he also took to task who criticized the *Ved* and *Purans*. He said :

वेद कतेब कहो मत झूठे। झूठा सो जो न विचारे।

*Do not call Veds and Purans as false. The one who does not understand the teachings of Veds and Purans is a liar or a fool.*

He is like a donkey who is loaded with sandalwood but remains oblivious of its aroma; such people have nothing to do with reality. Such people could be seen deliberately making unrestrained speeches everywhere.

Guru Nanak Dev established Sikh Sangat and Langar (community meal) in accordance with the fundamental philosophy of Bharat to end the then prevalent caste system in Hinduism. This establishment looks similar to the previous great men Bhagwan Buddh and Bhagwan Mahavir. I also believe that both these traditions *Sangat* and *Langar* have contributed a lot in the spread of Sikhism. The proclamation of *'Bole so Nihal'* ended the distortions by giving birth to collectivism, so that Hinduism could survive in and around Punjab. Guru Nanak Dev described the prevailing situation "The times are like a sword; rulers and kings are like slaughterers; goodness has vanished', love has blown away."

When Babur invaded Punjab in 1521, Nanak Dev had returned after visiting Arab countries and was staying with his beloved disciple Lalo at Syedpur, which is now Amnabad, 50 miles from Lahore. When Babur reached Syedpur and committed inhuman atrocities, the people of Syedpur resisted. Most of the people were slaughtered. All the young women were enslaved. Older women were forcibly employed to grind flour and cook food for the soldiers. The whole city was looted and set on fire. Nanak and Lalo also had to carry heavy items and take them to camps. They were also forced to run the mill. Nanak's tender heart was filled with compassion and pain. This intensifying pain floated out in the form of poetry :

> *You protected Khorasan and terrified the heart of Hindustan, but O Supreme Creator, why are you still uninterested? In the form of Mughal Babur, you have sent Yama yourself. The terrible massacre, the mournful high voices of the people have these not instilled a little kindness in your heart, O my Data.*

Describing the painful condition of the female captives, Nanak said :

> *"The beautiful head covered with hair, which used to have vermilion filled line, has now been shaved, locks have been cut ruthlessly with scissors., The daughters-in-law who lived happily in the palaces are begging on the streets. There is no place of refuge for them today.*
>
> *How beautiful these shaved heads were at the time of marriage! Sitting with their husbands in the ivory palanquins,*

*when these came to their home, the holy water was sprinkled on them in welcome.*

*The fans were oscillating around them. When she entered the house for the first time she received one lakh and one rupees as a gift. When she became a housewife, she again got one lakh rupees as gift. She used to eat dates and other soft fruits. That chair on which she sat became beautiful. Now they are tied with rope and pulled like animals.*

*Now, the neck garland is broken, pearls scattered, their beauty has become their worst enemy. The barbaric soldiers have made them captive and looted their honor.*

*A few, who escaped this barbarity came back to their homes. Other people would inquire from these escaped women about the condition of their loved ones. Not getting a satisfactory reply they would grieve. Some captives women would be lost forever, those who were saved, had to cry and suffer for the rest of their lives. Ahh Nanak! Man is helpless indeed".*

From the above poignant utterances, the pain of Nanak's heart and the inhuman atrocities committed by the invaders can easily be understood. At present, the purpose of this book is to show that Sikh sect, born out of the womb of *Sanatan Dharm*, also accepts the truth of spiritual harmony as the cardinal value of *Sanatan Dharm. Vedic* Rishis' proclamation एकम् सत् विप्राः बहुधा वदन्ति (truth is one, described differently by learned) was also accepted by Guru Nanak :

Guru Nanak gave the fundamental *Mantr,* which is described as below :

*Ik Onkaar Satnaam Kartaa Purakh Nirbhau Nirvair*
*Akal Moorat Ajooni Saibhang Gurprasaad*

What was already propounded by *Bhartiy Rishis* was confirmed by Guru Nanak Dev in the Guru *Granth* Sahib on page 1280. Guru Nanak Dev has said:

पर विरति निरविरति हाठा दोवै विचि धरमु फिरे रैबारिया।

*That is, in the eyes of Guru Nanak Dev, Dharm is the thread to tie both ends of tendencies and quietism and gave acceptance only to four Purusharthas i.e. Dharm, Arth, Kama, Moksh.*

Dr. Strump has also considered Guru Nanak Dev as an exponent of Hindu philosophy. Accepting this opinion,

Dr. P. Barthwal has written, “Guru Nanak has protected Hindu *Dharm* from the rising flood of Islam, as messengers of Hindu renaissance like Ram Mohan Roy, Ram Krishan etc. did. Hindu *Dharm* had to be a protected against the spread of Islam. Till the end, Nanak's style of thought remained completely in sync with Hindu thinking. He had contacts with many Muslims. A few Muslims also became his disciples. Those Muslims who became his followers were followers of Sufism. Sufism is nothing but Sarveshwar Vaad, which came directly from Hinduism. Its links with Islam are only external.” (*The Nirguna Sampradaya in Hindi Poetry,* page 66)

In *Bhartiy* thought, *Satyam Shivam Sundaram* has been considered as the foundation of *Bhartiy* culture. This principle has also been accepted in Sikhism through the concept of *Shubhi Karm* (good deeds). Its message is to embrace death with a smile, considering it to be filled with beautiful essence. A human being is the manifestation of imperceptible Bhagwan's Leela (God's play). Death merges the manifest with the unmanifest. This sublime feeling is the hallmark of the Sikh sect. Because of this truth, Guru Arjun Dev laughingly endured the tortures of Muslim invaders and the smile on his face remained the same even after he was made to sit on a hot pedestal. Finally, Guru Arjun Dev breathed his last. Guru Tegh Bahadur, laughingly got his head severed in Delhi. Due to belief in *Satyam Shivam Sundaram*, Guru Gobind Singh saw his two children torn into pieces and the other two being walled alive. These incidents could not disturb Guru Gobind Singh.

The ultimate goal of life is *Moksh*, that is, the union of the soul with the Supreme Soul. This essence of Hindu philosophy is the basic principle of Sikh philosophy too. Therefore, Sikhism is a version of Hinduism, purged of caste-based discrimination and rituals. Emphasis is placed on adopting an impartial vision of life.

The great message of 'सर्वे भवन्तु सुखिनः, सर्वे सन्तु निरामया' (everyone should be happy, everyone should be without illness) is the epitome of the eminence of *Bhartiy* philosophy and can be seen extensively in Sikh philosophy. It represents the highest levels of universal brotherhood. By giving the *Mantr* of social harmony and paving the way for the development of Rashtriyata, the Sikh sect rejuvenates dry and inactive philosophical thinking and makes it brilliant. Away from orthodox thoughts and by being imbued with the spirit of 'आत्मवत् सर्वभूतेषु' (everyone is like me) it opens the door to all human qualities.

The sacrifice of the Gurus and followers of the Sikh sect for the protection of *Hindutav* cannot and should not be forgotten. We must appreciate proclamations of Guru Gobind Singh : सकल जगत में खालसा पंथ गाजे। जगे धर्म हिन्दू सकल भन्ड भाजे।। [May *Khalsa Panth* (the path of truth) be in whole of the world. May Hindu *Dharm* get awakened and all frivolities vanish]

The resplendent *Nakshatr*, the tenth Guru of Sikhs of Bharat, Guru Gobind Singh Ji, is a great exponent of *Hindutav*. For the ignorant and people with the anti-Hindu mentality, the above proclamation is enough to tell them that Sikh sect is not separate from Hinduism. In critical times it descended to protect Hindu *Dharm*, to articulate the fundamental principles of *Hindutav* : the meaning of the creation of caste and *Varn*-less society as opposed to the *Varn System* based on birth. Many Gurus, from Guru Nanak Dev to Guru Gobind Singh, have clearly propounded the *Varn System* on the basis of attributes and *Karm*.

It is clear from the above discussion that like many other sects, Sikh or *Khalsa Panth* was also derived from Hindu *Dharm*. The philosophy of Sikhism is the expression of *Vedic* philosophy. The sacrifices made by the followers and Gurus of the Sikh sect to protect the Hindu *Dharm* are a clear historical evidence of the fact that Sikhism is an off-shoot of Hindu *Dharm*. In history books, these sacrifices are written in golden words. Those people who are bent on separating the Sikh sect from the Hindu *Dharm* by forgetting the spirit of incarnated Gurus and their sacrifice deserve to be strongly condemned. In fact, they are neither true devotees of Gurus nor well-wishers of Hinduism.

I want to question those who represent Sikhism as separate from Hinduism as to why did the followers of Islam make Guru Arjun Dev sit on a burning hot pedestal? Why was he killed through inhuman torture? Why did Kashmiri Pandits go to Guru Tegh Bahadur for their protection when they were forced to convert to Islam by Sher Afghan Khan, the then ruler of

Kashmir on the instructions of Aurangzeb? And why did Guru Tegh Bahadur got worried and thought that the sacrifice of a great man was needed to protect Hindu *Dharm*? Seeing his worried father, a nine-year-old Guru Gobind Singh said :

तुरकन भार दुखत भई लोई। छत्री जगतन दिखिअत कोई।।
जो निज अपनो शीश चढ़ावै। निधरन धरन भार ढ़हरावे।।
तुम में और अधक को आही। देग तेग जाके गृह माही।।

गुरु विलास 5, 15–16

*The atrocities of the invaders are increasing. Can some Kshtriy be found who can sacrifice his head for the protection of these poor people? In my opinion expect Guru Teg Bahadur no one else is fit for this purpose.*

Guru Teg Bahadur, the worried but determined religious protector sent the Kashmiri Pandits back, saying, "Send a message to Aurangzeb that if our patron Guru Tegh Bahadur accepts Islam, then Kashmiri Pandits will also accept Islam" Guru Teg Bahadur said so because he himself was a Hindu and wanted to save Hindu Dharm. When this news reached Aurangzeb, he summoned Guru Tegh Bahadur. Guru Tegh Bahadur fearlessly went to Delhi but was arrested midway.

To accept Islam, various temptations were given to Guru Tegh Bahadur by Aurangzeb. When temptations could not lure him, Aurangzeb started torturing Guru Teg Bahadur. On seeing Guru Tegh Bahadur steadfast and unmoving like the Himalayas, his disciple Matidas, who had accompanied him, was cut into piece with a saw in front of Guru Tegh Bahadur. Disciple Matidas' body parts were hung at Chandni Chowk. Many disciples were boiled in a pan of boiling oil, but the great sage Guru Tegh Bahadur continued to bear everything fearlessly. Eventually the cruel demon Aurangzeb ordered the beheading of Guru Tegh Bahadur. On November 11, 1675, he was beheaded at Chandni Chowk in Delhi. Guru Teg Bahadur did not leave his Hindu *Dharm*. For him self-sacrifice for the sake of *Dharm* was preferable. Let us try to understand the reasons behind this.

My question is, why was his disciple Bhai Matidas, ripped with a saw and his limbs hung in Chandni Chowk? Why was the disciple Bhai Dayaldas boiled in a pan of boiling oil? Guru Gobind Singh has commented on this sacrifice :

तिलक जंजु राखा प्रभुताका। लीनो बड़ो कलू महि साका।
साधनि हेतू जिमिकारी। सीस दिया पर सी न उच्चरी।।

विचित्र नाटक 5/13

*For protection of Tilak and Janeu (forehead mark and scared thread), Guru Tegh Bahadur gave his life but not his Dharm and there was no sign of pain or anguish in his eyes.*

Delhi's Gurudwara Sis Ganj is a loud testimony of this sacrifice. It is believed that a sudden strong storm engulfed the surrounding area when Guru Tegh Bahadur was beheaded. Taking advantage of this, his disciple Lakhi Shah took Guru ji's torso and hid it in his house so that Aurangzeb could not insult it. As arrangements could not be made to cremate Guru Sahib's torso, Lakhi Shah put his own house on fire. Today's Gurdwara Rkabganj is built at this place. Guru ji's head was taken to Anandpur (Punjab) by another disciple, Rangreta. His son Govind Singh duly cremated his father's head. This sacrifice shook the roots of Aurangzeb's rule. Guru Gobind Singh narrated this fact in the following way :

तेग बहादुर के चलत, भयो जगत में शोक।
है है है जग भयो, जै जै जै सुरलोक।। विचित्र नाटक 3,13

*The world mourned the death of Guru Teg Bahadur but it was celebrated in Surlok (abode of Gods) where Guru Teg Bahadur went after his death.*

After this, his son Guru Gobind Singh, took the command and occupied the Chair of Guru Tegh Bahadur to protect *Dharm.* He established the *Khalsa Panth* and formed an army, who had arms in one hand and scriptures in the another. In spite of not being powerful, this army, with sheer determination fought against Aurangzeb's resourceful army and dissipated all his power. Guru Gobind Singh famous saying is, 'सवालाख से एक लडाऊँ चिड़ियों से मैं बाज तुड़ाऊँ' (My one Sepoy will fight against 1.25 lakh enemies. My sparrows will break neck of eagles). This is a wonderful example of his heroic resolve and valour.

'तुर्क मलेच्छन सों नहीं मिलना, ले हथियार सामने पिलना' (do not befriend invaders and barbarians. Confront them with weapons). This lion like roar of great and fiery Guru Gobind Singh must have inspired millions of soldiers in that era. It is equally inevitable and inspirational in present times. To defend his own Hindu *Dharm* this brave and great man, with a lion's roar sacrificed his entire family. Sikh Gurus, especially Guru Gobind Singh, the founder of the *Hindu Pad Padshahi*, and Veer Shivaji had protected Hindu *Dharm*, otherwise the whole of Bharat would have accepted Islam like Arabia, Egypt, Indonesia etc. Prevailing terrorism and increasing population of Muslims is the new (modern) version of Aurangzeb's Islamization policy and is as deadly as Aurangzeb was. The conspiracy to separate Sikhism from Hinduism is also a part of their strategy.

Astu! a deep analysis of the history and scriptures of the Guru *Granth* Sahib and the Sikh Panth, makes it crystal clear that the Sikh Panth is a new version of age old *Sanatan Dharm*. It is natural that in the Guru *Granth* Sahib the acceptance is of the eternal principle of the *Varn System* as not based on birth, but determined by *Gun-Dharm*. It is visible in Guru *Granth* Sahib as well as in all other available texts of the Sikh.

# PART-II

# Distorted Form of System

For the smooth functioning of the social fabric the *Varn System* was established by the *Rishis* on the basis of attributes and *Karm*. *Varn System* had nothing to do with birth. In relation to the *Varn System*, the *Vedic* literature conceived the society as a whole body. The sun and the moon are the eyes of this body; the constellations and the sky are its navel and the *Brahmans* are its mouth, the *Kshtriys* are its arms, the *Vaishys* are its thighs and the *Shudrs* are its feet. These four *Varns* are four limbs of Hindu Society conceived as a whole body. It is clear from the following description about the Almighty in *Purushsukt*.

ब्राह्मणोंऽस्य मुखमासीद् बाहू राजन्यः कृतः।
ऊरु तदस्य यद्वैश्यः पदभ्यांशूद्रो अजायत।।

*His face became Brahman, his hands were made Kshtriy, his thighs became Vaishy and his feet became Shudr.*

Society, as the manifested form of Almighty Universal Being is our sacred and original concept of Hindu and *Bhartiy* ideology.

With an ulterior motive both Muslim and British rulers made conceited and systematic effort to destroy this fundamental concept of *Varn System*. Without dividing the Hindu society it could not have been possible. Unfortunately the politicians, who took the reins of power of independent Bharat started working on the same policy of 'divide and rule'. Many distortions were mischievously incorporated in *Bhartiy* theology and history during the period of Muslim and British rulers. The purpose was to spread confusion in Hindu society so that Hindu society remains scattered and fragmented. Similar interpolations were also carried out in the history books. The rulers of independent Bharat used these interpolated verses or pages as weapons with the intention of dividing the Hindu society and making them their own puppets. The message of these interpolations is that the plight and untouchability of *Dalits*, their exploration and victimization is the result of Hindu scriptures composed by *Brahman* and Brahmincal society. It is completely a baseless and white lie.

Astu! Hindu scriptures consider the universe to be created by *Brahm* and the *Rigved*, *Samved*, *Yajurved* too have been descended from *Brahm*. This is the reason why every human being was addressed as *Brahman* in *Vedic* Literature. Subsequently, people started doing their work according to their ability and capabilities. This basis of *Gun-Dharm* gave birth to the *Varn System* in the same way as in present times four classes exist in government jobs namely Class I, Class II, Class III and Class IV. These four classes are also a form of *Varn System*. This classification is the crystallization of the sacred purpose of the *Varn System* as laid down by our *Rishis* to organize the society on the basis of distinctive attributes and *Karm*.

In relation to the fundamental purpose of the *Varn System*, sufficient light has been thrown in the chapter *'Varn Dharm'*. Therefore, it has to be understood that as per the *Bhartiy Varn System* a *Varn* is attained by a person through his own merits, *Karm*, self-study and hard work. These days by abusing the eternal *Varn System*, a disgusting game of proselytization is being played all around. People, funded by the sponsorers of this game are doing the work of spreading confusion in a planned manner. If malicious designs of these vicious people are not aggressively attacked by putting forth the facts of the original form of *Bhartiy* scriptures, the entire Hindu society will be destroyed. Since these propagators, funded by Christians or institutions formed with their funds and fundamentalist Islamic institution, born out of the spirit of Islamic imperialism, actively participate in this heinous conspiracy to harm Hindu Society. Such sold out intellectuals interpolated pages in Hindu Scriptures. This process has been going on for almost twenty centuries. Even today, such people are engaged in unrestrained babble on Hindu *Dharmik* scriptures.

It would be appropriate that by removing the spectacles of prejudice that one can systematically study the *Bhartiy* scriptures and try to understand its fundamental nature and spirit. Only then it would be possible for us to have a vision of the sublime form of *Bhartiy* theology. Macho perversity and obstinacy blinds a person to the extent that he cannot understand the truth and differentiate it from falsehood.

Astu! The truth is that there was no trace of untouchability in Bharat before Christ, nor was this perverted form of the *Varn System* in vogue. After the death of Jesus, churches came into existence. These churches committed innumerable inhuman atrocities on the whole world. A mere thought about these atrocities gives goose bumps. The atrocities inflicted on Jews shook the whole world. Bharat was not be affected to a great extent. Around 6th century, Islamic invasions began on Bharat. With a sword in one hand and a *Quran* in another, the massacre started on the instructions of the Khalifa. From here begins the era of distortion. To understand this properly, it is necessary to understand the principles of Islam i.e. the principles of *Quran*.

# Distortions in the Original Varn System

*Bhartiy Varn System* emerged from sacred thoughts, but subsequently distortions started taking place in it. The Islamic invaders who aspired to enslave Bharat, worked in a systematic manner towards turning this spark of distortions into a volcano. Britishers followed suit. To understand the origin of the distorted nature of the *Varn System*, we have to understand the Islamic and British mentality and their well-planned conspiracies. It is absolutely expedient to discuss the ideological background of Islam and Christianity.

**Origin and background of Islam**

1400 years ago in Mecca, Arabia, Islam originated from the last Navi Hazrat Muhammad Salahu Wasallam. The *Quran Majeed* was revealed to him. When Muhammad Sahab turned 40 years old, he became the Navi there. Since then, the *Aayats* of *Quran Majeed* began to be revealed to him, which continued for 23 years. Islamic scholars believe so.

Vengeance was the reason for the rise of Islam. In the beginning, Muhammad escaped to Medina, gathered power by residing and fighting with native people. Resultently, there was huge massacre in the wars. It is a strong proof of vengeance. There are many *Aayats* in the *Quran Majeed* which are full of description of resistance and bloodshed. That's why Arab, after the rise of Islam, adopted the policy of expansionism. The first target of Islam was the Aryan i.e. Iran. Along with Iran, Bharat was also included in the plan of expansionism. In fact, Islam spread itself on the strength of the sword. The truth is that even today Islamic terrorism is inspired by the *Aayats* of the *Quran*. Muslim invaders marched towards Bharat with huge armies and sword in one hand and *Quran* in another. The first attack was made on Thane, in 636 AD and in 637 AD. Sindh was attacked in 644 AD. In 712, Muhammad bin Qasim invaded Bharat. After the capture of Deval, all people above the age of seventeen were executed.

As per the instructions of *Quran* the children and women of the dead were enslaved as *Mal-e-Ganimat*. Temples were destroyed. Mosques were built in their places. After this, Muhammad bin Qasim attacked Raja Dahar. Unfortunately, Raja Dahar was defeated and killed. His defeat engulfed Bharat in misfortune and Islamic barbarism.

**Aayats and Islamic Rulers**

According to the latest researches, the period of the Gupta dynasty is called the golden period of Bharat. After the end of the Gupta period, Bharat fragmented as a *Rashtr*. Bharat, which had shattered the arrogance of an invader like Alexander, gradually moved towards the worst. Before proceeding further, I consider it necessary to clarify one more point. Sometimes we equate Greek invaders with Islamic invaders, which is not correct. History has addressed Greek invaders as Yavan, but unfortunately and out of confusion, some people also address Islamic invaders as Yavan. This is completely false and misleading. In spite of being foreign invaders, the Greeks were not Muslims. The Greek society was in sync with the *Vedic Bhartiy* society. It is possible that the *Rishis* of Bharat may have imparted the message of importance of *Vedic Bhartiy* Society to Greece. This is the reason why the Greeks used to perform *Yagy*, *Havan* etc. They also believed in Gods and Goddesses, whose names correspond to the Gods and Goddesses of Bharat. Due to the change of language, just the pronunciation of names have changed. It is true that Greeks were fond of learning and were civilized, whereas the Muslim invaders appear to be full of religious bigotry, wildness and destructive and demonic tendencies. The Yavans, out of greed had plundered Bharat's prosperity and Alexander also had the desire to be called great and Chakrvrti. Because of the sharp edge and valor of the swords of Bharat, Alexander and Greek invaders like Seleucus, Demetrius and Menander were not successful in complete enslavement of Bharat.

When the impact of the Greeks began to decline outside Bharat and within their own country, they collapsed. Emperor Pushymitr dragged and killed Greeks and drove them beyond the borders of Bharat. The Shaks of Central Asia destroyed the Greek Empire. After this, Islamic expansionism thrashed the Greeks and converted them to Islam. Bharat was equally strongly attacked by Shaks, but they were defeated. However, our amazing digestive power completely absorbed Shaks within Bharat. Similarly, after the Shaks, we completely destroyed the invasions of the Huns and Kushans and assimilated them too. Subsequently, began the bloody series of years of continuous attack, looting, rape, arson and kidnapping by Islamic robbers who were inspired by the *Aayats* of the *Quran*.

Though, after the end of the Gupta period, Bharat had fallen to other distortions too and was divided into small kingdoms, yet these small states and kingdoms fiercely fought with these bandits and human-demons. Unfortunately, Bharat having a culture of morality, had no element of inhuman and beastly conduct in it. Innocent brave sons of Bharat were caught between the excessive generosity of Bharat and ungrateful and deceitful acts of Muslim bandits. Along with the rise of Islam in Arabia came the expansionist policy, which was guided by the *Aayats* of the *Quran*. This is the reason why invasions on Bharat began under the directions of Muslim Khalifa. The attacking warlords were directly under his control. Khalifa bore the expenditure on the formation of armies and all the other related expenses. Khalifa had 2/3 share in the loot. Sometimes this part was up to 3/4$^{th}$. Army commanders were clearly instructed to act in accordance with the instructions of the *Quran*, therefore, the looting method of every Muslim invader was the same. These invaders were glorified because everything done by these invaders was mainly in accordance to the *Aayats* of the *Quran*. Every human being get mentally disturbed if he reads the list of heinous crimes and other exploits of Muslim invaders in Bharat. When I read this history, tears would not stop flowing from my eyes.

It is not the subject of this book to describe the crimes committed by invaders, still, as it relates to the topic of the book, I will discuss these briefly. I consider it my duty to discuss as much as is inevitable for this book.

Allama Iqbal is no more in the world. Before going to Pakistan, because of the purity of the land of Bharat he wrote -

'मजहब नहीं सिखाता आपस में वैर रखना' (Religion does not teach us to hate each other). Self-gratifying speakers and thinkers do not get tired in beating the drum of secularism with this stick. Had Iqbal read the history of three or four preceding generations, he could not have written this song. Did he know what injustices, atrocities his *Bhartiy* ancestors had to endure? Islamic invaders treated *Bhartiys* worse than animals. Their mothers and daughters were robbed and were sold like sheeps and goats in the Arabian markets. Females were punished for being beautiful. These Muslim invaders, like demons and wolves, clawed on their skin. They were forced to convert to Islam. Did Allama Iqbal know all this? Religious mosques and Madrasas were erected by destroying the centers of Hindu faith. Allama Iqbal, after his migration to Pakistan did what his co-religious believers did. After the partition of Bharat into two nations Allama Iqbal migrated to Pakistan and composed songs in praise of Pakistan. Well, the next line of the same song shows the sentiment in praise of Bharat by saying : 'हिन्दी हैं हम, वतन हैं हिन्दुस्ताँ हमारा' (We are all *Bhartiy* and Bharat is our land).

If we read these two lines together, then it is clear that religion does not teach hatred. According to the Hindu thought, only the eternal values of humanity are considered to be *Dharm*, and the cult or method of worship is not considered to be *Dharm*, It is, only and only, the result of the great and eternal tradition of *Bhartiy* thought.

In the beginning of his book the *Muslim Sultans in Bharat*, renowned historian Shri P.N. Oak has written, "Let us read that piece of *Bhartiy* history in the Middle Ages, in which gluttonous, superstitious Arabs, on the pretext of propagating Islam, trampled on earth and shed rivers of blood. These vagabonds, nomads and morally inferior people went everywhere, entered every house. They had a blood-soaked sword in one hand, a burning torch in the other. These masquerading goons, dragged women and children into adultery and slavery. This form of religion or caste is such a blot that its darkness defeats even the devil. Bharat was one of those countries which was badly burnt, torn apart, crushed, paralyzed and crippled; people were taken prisoners. Nonetheless, Bharat, dealt with them with utmost humanity. For thousands of years these dreadful invaders kept coming like ocean waves. These scoundrels kept coming till their last Muslim ruler was put to sleep in Rangoon's tomb in 1858 AD."

Islam itself was the sponsorer of these gruesome activities and attacks. The supreme religious leader was the Caliph sitting in Baghdad. Arabic historians themselves have also made this confession. According to the Persian historian Ahman Ibn Yahya-al-Baladhuri of Abbasid, who wrote in the *Futuh-al-Buldan* that the physical head of the religious headquarters of Damascus, the Khalifa, with the help of the deputy head of Baghdad (Iraq), planned these programs of plundering.

This fact has been stated by Swatantra Veer Savarkar, "In the end Osman, the governor of Oman, the chief Caliph of the Arabs, vehemently attacked Sindh. The then *Brahman* king Chacha of Sindh defeated his Arab army and killed the commander Abdul Aziz. After these battles and till 640 AD, these Arabs did not cause any significant disturbance. Only by conquering a remote, isolated small region called Markani, he forcefully converted the Hindus into Islam. These very people are Baloch people who later became staunch Muslims."

In the beginning of his book, Oak has described the inhuman and abominable acts done in Bharat by the followers of the *Quran*, "By entering Bharat these barbaric gangs, like termites and grasshoppers sucked this country. A grand country, where rivers of milk and honey flowed in royal palaces and picturesque buildings were adorned with pearls, gold and diamonds and were illuminated, was converted by these gangs into slums with open drains, huts and unbricked houses."

In the same book, attacking the sycophants and false historians, Oak has exposed the dark face of Islamic invaders, "In these circles, only drunken and opium-drenched debauchers occupy the thrones of polymorphs, of offering severed heads in bags, of minarets of cut off heads in mass massacres in every war and rebellion; where in *Harems* and brothels, lived thousands of enslaved men and women. These sensual rogues did unnatural sex and indulged in adultery; murder by demonic tortures; piercing the eyes; mass religious conversions under threat of rape, razors or hot bars; bribery; corruption; theft;

dacoity, and plundering of the wealth of Bharat and taking it to Arabia, Abyssinia, Iraq, Persia, Afghanistan and Turkey. They prohibited Hindus from riding horses, made them walk in their own homeland; forcing them to walk with a degrading colored stain on their clothes, treated their women and children as contemptible slaves. There is a description of kidnappings and thousands being sold as slaves, and the division of similarly confiscated property and human beings between the leaders of foreign tribes and their followers in a ratio of 1/5 or 4/5."

In the above context, in 6th volume of the *Savarkar Samagra* by Swatantra Veer Savarkar, at page no.140 it has been commented, "Amongst many of the wars waged by Muslims against Bharat, this one was unprecedented and was a two-pronged attack on Hindu *Rashtr*. All foreign invaders like Yavan, Shaks, Huns who invaded Bharat before Muslims had only one purpose i.e. establishment of their rule by conquering Bharat. Apart from this political longing, there was no other cultural or religious animosity behind their attacks. This new Islamic enemy, like other enemies had an ambition to establish a Muslim empire in the whole of Bharat by snatching power from Hindu *Rashtr*, but the ambition of religious conversions was not in the dreams of the earlier invaders. This burning religious ambition, constantly kept these invaders motivated, incited and instigated for their attacks on Bharat. These attacks were many times more monstrous and destructive than their political ambition. These invaders, on the strength of sword wanted to destroy *Hindutav*, the very basis of Hindu *Dharm* and Hindu *Rashtr*. Thousands and thousands of Muslim invaders from all over Asia scampered, pounced and fringed upon Bharat to impose Islam on Hindu *Rashtr*."

From these instances it is clear like a mirror that the followers of Islam did not come to invade Bharat just to establish their rule, but they also came with the determination to forcibly impose Islam on Bharat. To achieve it they resorted to inhuman, heinous and despicable acts. Crossing all limits of ruthlessness, they caused rivers of blood to flow. They were sexually perverted to such an extent that they forced noble ladies to lead a life worse than that of prostitutes. Guru Nanak Dev's comments on their condition are given in the chapter titled "References from Sikhism".

In fact, if we look at some *Aayats* of the *Quran* as a hallmark, then we clearly see the spirit contained in these. If we look at these *Aayats* in the mirror of history, we find that history is nothing but a reflection of these *Aayats*. These *Aayats* are given below :

*"Fight those who do not believe in Allah or in the dooms day and who do not consider unlawful what Allah and His Messenger have made unlawful and who do not adopt the religion of truth from those who were given the Scripture. Fight until they give the Jazya willingly while they are humbled." (Sura 9 Aayat 29)*

*"Those who disbelieve in our Aayats, we shall make them enter Fire soon; as often as their skins are thoroughly burned, we shall exchange them for other skins, that they may taste the torment; truly, Allah is All-Mighty, All-Wise." (Sura 4 Aayat 56)*

*"Momino take up arms for jihad and then go out as Jamaat-Jamaat. Do it."(Sura 4 Aayat 61)*

*"Those who wage war against Allah and His Messenger, and go about the earth spreading mischief, indeed their recompense is that they either be done to death, or be crucified, or have their hands and feet cut off from the opposite sides or be banished from the land." (Sura 5 Aayat 33)*

*"Those who do not give orders according to the orders of the Prophet of God, such people are infidels and we ordain for them therein a life for a life, an eye for an eye, a nose for a nose, an ear for an ear, a tooth for a tooth, and wounds for wounds, is legal retribution. But whoever gives up his right as charity, it is an expiation for him. And whoever does not judge by what Allah has revealed, it is those who are the wrongdoers i.e., the unjust." (Sura 5 Aayat 44, 45)*

*"Khuda wanted by his decree to uphold and protect your rights. Cut the roots of disbelievers and throw them away." (Sura 8 Aayat 7)*

*"And fight with them until there is no more persecution (fitnah) and religion should be only for Allah; but if they desist, surely Allah sees what they do." (Sura 8 Aayat 39)*

*"Eat enjoy, utilize of what spoils you have taken, what is lawful and good; and fear Allah, verily, Allah is forgiving and merciful." (Sura 8 Aayat 69)*

*"If you could only see when the angels took away the souls of the unbelievers, striking them on their faces and backs, saying: Taste the torment of burning." (Sura 8 Aayat 50)*

*"Therefore, if you overtake them in fighting, punish them in such a way that those who are supporting them from behind, leave them and run away." (Sura 8 Aayat 57)*

*"Rouse the believers to engage in battle. If there are twenty steadfast among you, they will defeat two hundred; and if there are a hundred of you, they will defeat a thousand of those who disbelieve; because they are a people who do not understand." (Sura 8 Aayat 65)*

*"It behoves not a Prophet to take captives until he has sufficiently suppressed the enemies in the land. Kill infidels and let earth be drenched in their blood." (Sura 8 Aayat 67)*

*"But when the forbidden four months are past, then fight and slay the pagans wherever you find them, and seize them, beleaguer them, and lie in wait for them in every stratagem (of war); but if they repent, and establish regular prayers and practice regular charity, then open the way for them." (Sura 9 Aayat 5)*

*"(A command will be issued): "Seize him and shackle him, then cast him in the Fire, then fasten him with a chain, seventy cubits long, and give no food except the filth from the washing of wounds, which only the sinners will eat." (Sura 69 Aayat 30-37)*

*"Believers, do not take your fathers and your brothers for your allies if they choose unbelief in preference to belief. Whosoever of you takes them as allies those are wrong-doers." (Sura 9 Aayat 23)*

*"O you who believe! Fight those of the disbelievers who are close to you." (Sura 9 Aayat 123)*

*"Fight against the disbelievers and the hypocrites and be harsh upon them." (Sura 9 Aayat 73)*

This is only an indication. It is not possible to cite all *Aayats* of the *Quran* in this book. It is not even necessary for the present subject of the book. This barbarism was instrumental in bringing the *Bhartiy Varn System* to untouchability. That is why it is necessary to present this inherent core spirit of Islam. That is why I have mentioned these *Aayats* as a hallmark. I have, many times, read the *Quran* published by Mahmood & Co. Marol, Pipeline 56 and translated into Hindi by Maulana Fateh Muhammad Khan Sahib Jalandhri from beginning to end. After reading it, I have come to the conclusion that the blind follower of the *Quran* can never be a tolerant being.

Many such *Aayats* of the *Quran* were followed completely by the Muslim invading robbers. The following similarities are seen in every Muslim invader or ruler -

---- They were either employees or were acting under the directions of Khalifa or were inspired by *Aayats* of the *Quran*.

---- They not only wanted to establish their own kingdom over Bharat, but their intention was also to force Hindus to convert to Islam.

---- They gave gifts to the converts. They either beheaded those who did not accept Islam or they made them slaves and sold them like animals in the Arab markets or were made untouchables and were separated from the rest of the society.

---- They were fanatic, ruthless and barbaric.

---- They were lustful, characterless and sexually perverted. They were also addicted to most unnatural sex. That is why along with women and girls, they used to kidnap boys too.

---- Their style was that of plundering. According to the *Quran* they robbed women, girls, boys of their property, considering these to be *Mal-e-Ganimat,* whose right of consumption was given to them by their religion.

---- Whichever city they attacked, they targeted the temples first. Breaking idols, looting them, converting temples into mosque were their aims. With minor modifications or architectural changes minarets, platforms of temples were changed into mosques. In order to target the idols of the reverence and Hindu faith, they used to break them and get them installed on the steps of the mosques.

---- Their natural instinct was of deceit, stealth, bolony and guile.

---- They used converted Hindus to cause atrocities on the rest of the Hindus. There policy was to use blood against their own blood.

---- Mass genocide, burning village after village, pouring poison into ponds and wells, demolition, breaking bridges, destroying standing crops, setting fire to barns, etc. were their forte.

---- For their carnal desires innumerable maidens, ladies and boys were always available in their *Haremes*.

---- Killing their own father, brothers and any other blood relatives was their peculiar characterstic.

Astu ! Despite the above wild characteristics, the history taught in schools in independent Bharat seems to eulogize invading rulers. Unfortunately, instead of showing the real face of these brutish ferocious vampires to our present and future generations we are hell-bent on proving them as Gods. What could be the real reasons? Well, anyone can guess.

It seems necessary to give a brief description of some of other Muslim invaders. As I said earlier, Islam, which was born in Arabia after the Greek invasions, forcibly converted central Asia, which had Hindu learnings, into Islam. After religious conversions, under the direction of the Khalifa, Islamic bandits, turned towards Bharat with a sword in one hand and the *Quran* in another. Khalifa Umar first invaded Bharat in 636 AD. The name of the chief of the war-gang he sent to Bharat was also Umar. This attack was carried out at Thane near Mumbai. But the brave hearts of Bharat mashed all of them into dust. After this, the same Caliph sent another gang of robbers. All these were buried in the ground. The third gang of robbers led by Mudheera was sent to attack from the north. By resorting to deceit, debauchery, lies and immorality this gang succeeded to some extent in penetrating the security of Bharat. Thus, this gang was able to reach Devyalpur. Devyalpur is now known as Karachi. Previously there was a huge domed temple of the deity. Devyalpur was named after this deity. According to the Arabic historian Al-Baladhuri, "The enemy was completely eliminated from this city and the temple was demolished and a mosque was built." Whereas P.N. Oak and many other historians call this description false. According to him, the invaders were made to bite dust by the *Bhartiy* fighters.

After Osman became the Khalifa, he appointed Abdullah as the ruler of Iraq. The Khalifa ordered him to send an espionage team to Bharat. Under his leadership, a group of spies came to Bharat. Abdullah was taken prisoner and punished by the guards of Bharat. He lost his mental balance. After his return to Baghdad, he described Bharat in such a way to Khalifa that the Khalifa could not muster courage to attack again.

Subsequently, Ali became the Khalifa. To attack Bharat in 656 AD, he sent a huge gang under the leadership of Abdi. According to historians, "Abdi was victorious. He plundered, imprisoned the people and in a single day beheaded thousands of Hindus. Leaving behind his few companions, he, along with all his other companions fled towards Kikan of Khorasan, (on the border near Sindh) and was killed in 662 AD."

After the invasion of Abdi, the Khalifa Muawiyah again took the help of another robber Mohallab. Under his leadership a large gang was sent to Bharat. He too was put to death by the soldiers of Bharat. After that, the Khalifa, with the help of the ruler of Baghdad, sent a bandit named Abdullah as the chieftain of a huge gang to Bharat. The brave *Bhartiys* of *Kikad* severely beat him and in order to save his life he ran away.

Giving great allurements, the Khalifa again got Abdullah ready for the attack. Again he could not successfully attack Bharat. He was killed and buried within the borders of Bharat. Similarly, Sinan, Ziyad, Manjar alias Abul Ashas, Ibnghari Albwali, Muzza and other chieftains attacked Bharat. Though, all of them did the work of burning Hindu houses, kidnapping Hindu women, demolishing temples, forcibly converting Hindus, enslaving and selling them like animals in Arab markets. Yet their attacks could not be very effective. They were either put to death or forced to flee by the shining edge of the sword of Bharat.

For almost 75 years these robber gangs, hovering over the border, shed blood of unarmed civilians, abducted helpless women and children, committed fornication with them or drove them like sheep and sold them in the markets of Baghdad and Damascus, burned village after village, killed and ate cows and demolished temples. They considered it their religious duty to convert Hindus into Islam.

*Bhartiy* soldiers defeated these invaders, but did not annihilate them. History would have been different, had the *Bhartiy* heroes followed the Chanky policy of exterminating the enemy. If the *Bhartiy* heroes had raged and killed them there would not have been such a long series of plunder and atrocities. Bharat could never be enslaved. Unfortunately, we still are not willing to learn from the mistakes of history. This is the reason, why till date we are standing on the verge of danger.

The above robbers forcibly converted our own brethren to Islam and caused their alienation from us. Invaders continually tarnished the glory of grand temple called Bharat. Instead of hating our own brethren, who were forcibly converted to Islam had we loved and assimilated them like Shaks, Huns and Kushans, invaders could not have found a ladder to defile the flag of our glory. With his fifty thousand soldiers and under the directions of Khalifa, a 17 years old rowdy, riotous savage, devil Mohammad bin Qasim used this very ladder to attack Sindh in 711 AD. The horror of this attack has been described by historian P.N. Oak on page 27 of the *Muslim Sultan 1 in Bharat* in this way : "This 17 years old devil created a storm of tyranny. One lakh Hindu women were imprisoned, seventy rulers (*Ranas*) of Sindh fell down, temples were converted into mosques by building minarets and platforms. The whole of Sindh became barren due to the transgression of arson and looting."

Drawing a picture of Islamic invasions, on the same page, Oak said, "The demonic web of plundering, burning, persecution, kidnapping, conversion to Islam, lustfulness and enslavement were executed in two ways in Bharat by the barbaric ungrateful Arabs. On the one hand, barbaric Arab gangs, equipped with horses, spears, swords, bows, arrows and drugs were sent to Bharat and on the other hand the sinful harvest was sent to the markets of Damascus and Baghdad. There were heaps of kidnapped Hindu women and children, looted gold, silver bricks, jewels, blood-stained heads of Hindu chieftains, disfigured idols and looted temple-treasures.

The Khalifa was the head of this management. This system of attacks was run by him. His accomplice was the ruler of Baghdad, who, with the intention of plundering, hovered over Bharat and used to send sinful harvest back home.

The bones of helpless Hindu women, innocent children and other human beings were scattered on the road from Karachi to Baghdad and Damascus. Inhuman tortures took their lives. Animal lust, bloody atrocities and extreme tortures shattered them. Many branches and trails emerged from this route. In the houses and buildings located on these trails lay scattered the looted wealth of Bharat."

Describing the civilization and misconduct of the Khalifa, Elliot and Dawson wrote on page no. 439 of their book, *Before the Sindh Conquest*, "We find that the followers of Khalifa filled the corpse of a donkey with the dead body of an Egyptian ruler and burnt it to ashes. Suleiman was the Khalifa when Moses conquered Spain. The cruel vampire, who had killed the Sindh conqueror, drove Moses out of his country.

Moses spent his troubled days in Mecca. Suleiman got his son murdered in Kerdova and threw his son's severed head at the feet of Moses. The followers of this vampire laughed and taunted the father, who was already in despair and anguish."

Regarding their cruelty and brutishness, Pascal de Gnosis has said in the essay titled "Alhajjaj" of the *Biographical Dictionary*, "It is said that this mad human-vampire through his followers got one lakh and twenty thousand people killed. After his death, about thirty thousand men and twenty thousand women were found in many of his prisons. This conclusion is from a Parsi source."

H.M. Elliot has written on page 433 of his book, "These cruel fanatics openly honed their lustful life in luxury and sensuality and propagated this type of religion all around."

At page 30 of the book *Muslim Sultan in Bharat* historian Oak wrote, "What a wonderful civilization that promotes treachery, plunder, theft, arson, rape, unnatural sex, destruction and genocide! It takes pride in converting temples into mosques, killing people and converting them into Islam."

P.N. Oak has written on page 26 of the same book, "This tradition of Muslim robbers of dividing the *Bhartiy* plunder into 2/4 and 4/5 parts continued till the end of Muslim rule in Bharat."

Ahh! The saga of incorrigible predators is being taught in a glorified way to our children in schools even today. When will the day come when the gruesome and realistic depictions of these dastardly predators will be brought to the minds of the future generation of Bharat and they will realize the truth! Only then, the revival of Bharat is possible. Only then, those wolves who are calling themselves intellectuals and pioneers just by killing Bharat's ancient intellectual property in the name of pseudo-secularism can be taught a lesson. I wish they had read *Aayats* of the *Quran*. Inspired by these *Aayats*, the inhumane atrocities on Hindus have been committed since the 6$^{th}$ century. Had we done a fair review of the dastardly acts, Bharat would neither have been moving again towards slavery and nor would have *Hindutav* reached the point of defenselessness.

There was no dearth of traitors during the reign of King Dahar of Sindh. At the time when Muhammad bin Qasim attacked, the Buddhists of Bharat unitedly supported Muhammad bin Qasim. King Dahar had also recruited a few Muslims in his army. On the one hand the Buddhists welcomed Muhammad bin Qasim and on the other hand Muslim soldiers of Dahar's army revolted on the pretext of religion. They made it clear that they will support Muhammad bin Qasim as he was a Muslim and a crusader who had come to Bharat to wage war in the name of religion. Consequently all Muslim soldiers attacked King Dahar's army.

The kingdom of King Dahar was divided into four segments. The first segment was Nirun, Devalayapur (Karachi), Lohana, Lakha and Samma. Its center was in Brahmanwad. In the second part, whose headquarter was *Shivsthan*, were the hilly areas of Budhapur, Janakan, Rajkan. The third part was of Talvadan and Chachapur i.e. Akshalanda and Pabia. The fourth region was Brahmpur, Karur, Ashahar and Kumb. Its headquarter was Multan. Alor was the capital of the kingdom of Dahar. The boundaries of the fourth part extended up to Kashmir.

As King Dahar was a very powerful and just king, misadventures of human-vampire were of no avail. Observing this fact, Khalifa finalized the plan with the ruler Hajjaj. Hajjaj appointed a 17 year old Muhammad bin Qasim, his brother-in-law and son-in-law, as the head of an army of robbers. The number of the soldiers of the gang was more than fifty thousands and they had machines to shower stones etc. from afar. The first war which Qasim fought against King Dahar was in Brahmanwad. Son's of Bharat fought bravely. Due to the news of King Dahar's killing in this battle, a stampede occured in the winning army.

According to some historians, King Dahar had two queens. One of those queens, Rani Mainawai with her fifteen thousand soldiers fought the enemy. She was defeated. The queen along with all the family members and women, committed Jauhar by jumping into the fire. According to some historians, King Dahar had two daughters Suryadevi and Parmaldevi and a younger queen Lado. These three were entrapped by Qasim. Qasim kept the queen Lado for his adulterous ventures and sent both the daughters to Baghdad for the enjoyment of Khalifa. But according to some historians, no member of King Dahar's family was captured by Qasim. The two young women, known as Suryadevi and Parmaldevi, who were sent to Baghdad for Khalifa, were, in fact, the daughters of some military officers. Be that as it may, these girls acted in front of Khalifa as daughters of King Dahar. I bow down to the two brave fighters of Bharat, who taught a lesson to these cannibal, sensual human-vampires. Their names have become immortal in history.

When Suryadevi and Parmaldevi were sent to the Khalifa in Baghdad, the Khalifa ordered for the elder Suryadevi to be decorated and presented before him. When the elder girl Suryadevi, whose real name was Janaki, appeared in front of the Khalifa, the sexually aroused Khalifa proceeded to take her in his arms. Suryadevi stepped back and said that as he was a great preacher of religion, he should not accept used things. Qasim had sent both of us to you after having enjoyed us for three days and Khalifa does not deserve them anymore. Hearing this, the Khalifa went mad. He apprehended Hajjaj, the ruler of Baghdad and said that his son-in-law Qasim must have had his blessings to do such a thing. Having said this, Khalifa took Hajjaj as a prisoner and gave an order to his loyal soldiers that Muhammad bin Qasim should be captured and brought wrapped and stitched before him and in the fresh skin of a bull. Loyal soldiers came to Bharat. They captured Muhammad bin Qasim, tied his hands and feet and sewed him in raw leather of a bull. He was brought to Damascus by keeping him in a box. On receiving this information, the Khalifa called the two *Bhartiy* girls and other courtiers. The box was opened in front of them. Qasim, sewn in raw leather, had died of suffocation. The human-vampire who wreaked havoc in Bharat was lying at the feet of those girls. They had avenged the killings of innocent people of their country. These girls were not satisfied with just this much. Right now the wolf Caliph was in front of them. Those girls laughed and said, "You are a fool. We took revenge for destroying our country. Qasim did nothing to us."

Elliot and Dawson on page 211 of their book have recorded the words of these two girls to the Khalifa in the following way, "Surely your command has been fulfilled, but your mind is completely devoid of justice, conscience and simple understanding. Qasim didn't even touch us, but that devil had killed our king, ruined our country, destroyed our honor and pushed us into the quagmire of slavery. That is why, for vengeance and revenge, we resorted to false story. He defiled twenty thousand women like us, put sixty rulers to death. He turned temples into mosques, minarets and speech platforms."

The Khalifa was stunned to hear the words of the girls. According to historians, he chewed his own palm in anger. He died of this stroke in January 715. Hajjaj was so shocked by this painful death of his brother and son-in-law Qasim that he died in June 714, six months after the incident. The Khalifa put those brave girls in prison. Khalifa Suleiman did not even dare to touch these girls. He ordered both of them to be tied behind a horse and dragged to death. The bodies of these two brave daughters of Bharat got scattered like rags on the streets of Damascus. These two girls became immortal as an example of valor.

Before the above incident, for three years Qasim continued to cause destruction in Sindh. He not only looted but he also destroyed many cities including Alor, Devalayapur

(Karachi), Brahmanwad, Budhia, Nirun, Shirsham, *Shivsthan*, Nilham, Shailaj, Bahitlur, Kanthvel, Bait Sagar, Raver, Narayani, Jaipur, Kajijat, Bahrur, Dahleela, Chanir, Batia, Jalwati, Multan, Mahal Sambandhi, Danda Karaha, Wahrawar, Lohana, Sihra, Brahmapur, Ajtahad, Karur, Rori and Udhampur.

By the might of sword, severe beating, imposing life and property tax on Hindus, Qasim converted people to Islam. The already plundered Hindus were devastated by the tax as these taxes were not imposed on converted Muslims. The intention was that everyone should become Muslim. Considering this as a religious act, an Arabic historian wrote, "He imposed a terrible tax on the basis of the law of the Prophet, and those who became Muslims were freed from slavery, property tax and life tax. For the rest of the people, whose houses were already badly robbed, heavy taxes were collected according to their past status."

On page 55 in the *Muslim Sultans in Bharat 1*, Oak wrote, "Qasim's historical hooliganism turned a sparkling province full of fields, idyllic lakes into a desert. A large section of the population was swayed away from their country and were converted to Islam. Cities and forts were reduced to ashes. Temples were turned into mosques."

Even after the martyrdom of King Dahar, Qasim had to taste the dust at many places. Both sons of King Dahar also fought fiercely and attained martyrdom. King Dahar's defeat and Qasim's victory actually appear to be standing on the foundation of deceit, guile and debauchery. It is not possible to describe in detail the massacres and genocide committed by Qasim in Bharat for three years. The horrific acts of genocide, plunder, rape, arson and kidnapping were done in accordance with dictates of the *Quran*. These acts put humanity to shame. Humanity gets swamped. All this in the name of Allah! Oh Religion! What a religion!

When the consignment of loot, along with women and children reached Hajjaj, the later sent a letter to Qasim, clearly ordering the brute acts to be done in the name of religion, "Don't give any chance to the infidels, behead them at once..... This is the order of Allah." In the same letter it was also written, "The

Khalifa has spent sixty thousand Dirham on this project. As promised, he has to give double of it to the Khalifa."

This letter has been mentioned by H.M. Elliot on pages 206 and 207 of his book. Hajjaj also wrote in this letter, "Wherever there are ancient palaces, towns, cities, there should be a mosque, minaret and *Azaan* platform and read *Khudwa*."

In these orders, the directions of the *Aayats* of the *Quran* are clearly visible. The gruesome, disgusting face and principles of Islam are very volatile.

Regarding the Arab robbers, rest of the people of Bharat were ordered, "If these robbers come to the houses of the remaining Hindus, the healthy guests should be entertained for one day and night and sick guest for three days and three nights."

It simply means that the self-proclaimed custodians of religion did the heinous act of making every Hindu house a brothel. Today, when I see someone defying Hinduism and advocating the liberalism of Islam, I see the blood of these human-demons flowing in their veins. Their hypocrisy is clearly understandable.

I want to tell to these traitors and hypocrites of our country that, had they read history, their delirium would have cracked. They would have realized that every invader tore humanity into piece and reduced it to fragments. All this was done in the name of Allah as given in the *Quran*. Why should our own traitors and hypocrites call these people great? Though Bharat had to undergo inhumane and devilish tortures yet two brave daughters, in their prudence, taught a great lesson to the invaders and their masters. These Arabian wolves, if happened to remember this incident used to get shocks.

This is the reason why after the painful end of Muhammad bin Qasim, for about 250 years these wolves could not muster courage to even glance at Bharat. Though suffering horrific genocide and brute tortures, the heroes of Bharat recaptured the land of Bharat, but did not take back our brethren who were converted to Islam into their fold, though these converted Muslims hated Islam as they had suffered injustice and atrocities and had seen all the horrific acts with their own eyes.

As a result, these converted Muslims became the right hands of aliens and started beheading their own people.

Astu! I wish our ancestors had assimilated them like the Shaks, Huns and Kushans. Had this assimilation taken place Bharat would neither had to suffer long years of slavery and nor would such a number of Muslims had grown in Bharat. This growth is no less than like a deadly disease for Bharat.

From 961 AD to 969 AD, Alaptagin, the ruler of Khorasan province, through his Turkish general, plundered the areas along with *Bhartiy* borders. Khorasan was under Samanid. The kings of Samanid were *Kshtriy*, but the poison of Islam destroyed their *Hindutav* and made them Muslims. After Turkistan, every bit of Hindu-Afghanistan gradually got absorbed in the stomach of Islam. Before this, Hindu countries like Iran, Iraq, Arvasthan etc. were devoured by Islam. Now the neo-Muslims, the sons of Mother Bharat, started sucking the blood of their own mother. The Turkish general of Khorasan, under the direction of such neo-Muslims, was also doing the work of scrapping the borders of Bharat, burning crops, kidnapping the helpless women and screaming children and sold them as slaves to neo-Muslim countries.

At that time, Jaipal was the ruler of Punjab and a part of Afghanistan. He landed trouble as instead of direct fight with armies, Subuktagin suddenly engaged in attacking villages for plundering, burning standing crops and kidnapping women and children. He had also captured Lamadhan. For 20 years, he made the surrounding areas completely impoverished. Later, this robber returned to Balkh. His eldest son was Mahmood Ghaznavi. Since he was brought up by his robber father, he grew into dreadful man who did not hesitate in insulting his father. So Subuktagin made his younger son Ismail his successor. In Ghazni, Mahmood imprisoned his brother Ismail and at the age of thirty became the chief of the dacoit gang. Gradually he acquired all rights over Ghazni. Because of the deep marks of smallpox on his face Mahmood was very ugly and equally cruel. Because of his upbringing and teachings of Islam he grew cruel as a wolf. He has been portrayed by Islamic historians as a great saviour and architect of literature and art, though his character was no less than that of a barbaric tyrannical monster.

Regarding Mahmood Ghazni, Muhammad Habib, professor of Aligarh Muslim University, in his book the *Sultan Mahmood of Ghazni*, refutes the above claim and has written, "It was because of utter greed for power that he attacked Bharat. The life of the Sultan clearly shows that, whatever he may have been, he was never the ideal form of good qualities, as the fanatical Muslims portrayed him to be. In matters of wine, women and in battle his moral character was similar to that of the previous invaders. In order to keep the Turkish slaves subjugated, he used to snatch his subordinate officers. He also had many illicit children."

With regard to Mahmood, Dr. Edward Sachu has written, "All Hindus are infidels for Mahmood. They all deserve to be sent to hell, because they refuse to be looted."

What's more, Mahmood's own salaried writer Alberuni has written, "Mahmood annihilated the progress of the country. As in the stories of grandmother, he tried to show such miracles that Hindus were shattered to pieces. Their scattered pieces have given rise to a tendency to hate Muslims that will never end. That is why they relocated centers of knowledge and science to places like Kashmir, Banaras etc., far away from the regions which we have conquered. For political and religious reasons, the enmity among foreigners and *Bhartiy* has been increasing."

It is clear from these examples that Mahmood Ghaznavi was a religious fanatic and animalistic. He imprisoned his brother Ismail and crafted a rift in the family of king of Samanid to whom he was a subordinate. One of his brothers was killed by him and the other brother joined him. He divided the area of plunder into two parts. He kept one part with him and established independent power. Seeing the growing power of Mahmood, the Khalifa showered his blessings and sent him a pure garment and bestowed him with the title of Sultan Amin-ul-Millat Yamin-ud-Daulah. Now he was in Khalifa's good books. Prof. Habib has written on page 23 of his book, "Mahmood Ghaznavi pledged that every year he would launch jihad on Hindus. Out of 30 years of his life of robbery he attacked Hindus seventeen times. Therefore it is true that he fulfilled his promise hundred percent."

This comment of Prof. Habib supports our point that Mahmood wreaked havoc on Hindus, committed inhumane atrocities, converted them forcibly, demolished Hindu temples and idols, raped and sold women and children as slaves. All this was either religious hysteria or it was the result of the education given by the *Quran* or he acted at the behest of the Khalifas.

In 1000 AD, along with a squad he again attacked Bharat. Crossing the Indus river, he wreaked such havoc that the whole area became a desert. He caused great bloodshed. Only those who had accepted Islam were spared. All Hindu temples were converted into mosques. He returned to Ghazni with huge wealth in the form of enslaved women and children.

From 1000 AD to 1030, the human-vampire Mahmood carried out a total of seventeen attacks on Punjab, Delhi, Mathura, some areas of Rajasthan, Gwalior, Kalinjar, Varanasi and Somnath, etc. In these attacks, about 20 lakhs Hindus were killed, thousands of temples were converted in to mosques. Lakhs and lakhs were converted to Islam. Property, worth billions was looted and sent to Ghazni. He destroyed Hindu temples as per the program of a fanatic Muslim invasion. Burning village after village; killing Hindus as cruelly as was possible; kidnapping hundreds of women and looting wealth, he kept returning to Ghazni. After regaining strength, like bandits, he would come back to repeat the same orgy. The temples of Somnath, Mathura and Banaras were destroyed by him. Due to his religious bigotry he broke the Shivling of Somnath temple into pieces. Not only this, he got a part of this Shivling laid on the steps of Jama Masjid of Ghazni, so that every visiting Muslim would crush the Shivling with shoes, the other part of Shivling was buried in Ghazni's horse racing ground. His aim was to loot the accumulated wealth of temples. Apart from this, he wanted to exterminate all Hindus in every possible way. He had a well thought off strategy to destroy the centers of Hindu faith. The Khalifa was extremely happy at his actions. He bestowed on him the title of Butshikan (destroyer of idols). At this juncture, I would like to point out the pledge undertaken by Mahmood and described by Prof. Habib in his book. In the open court and in front of the Khalifa of Islam, Mahmood had vowed, "Only after annihilating all the infidels of Bharat, will I call myself Sultan."

Prof. Habib has further written in this book, "Mahmood was rolling in infinite wealth. *Bhartiys* started hating his religion. The looted people will never look at Islam in a congenial way. He has left behind the ever-living stories of robbed temples, ruined cities and crushed corpses. This has only led to the moral decline of Islam as a religion, let alone raising the moral standards, its loot has been estimated at three million Dihram."

On page 76 of the *Muslim Sultans in Bharat Part-I*, P.N. Oak wrote, "Thousands and thousands of ordinary helpless *Bhartiy* farmers, women, children were dragged to Ghazni. Their value was two-three Dihram in the markets. Therefore, in addition to the loot of gold seals, gold and silver bricks, gems, jewelry, thousands of *Bhartiy* prisoners were sold in slave markets and out of these sales he made several millions (one million is equal to ten lakhs)."

After amassing incredible loot, the news of robber Mahmood's return to Ghazni used to spread with the speed of light. Flocks of Muslims, from far-off places like Bhavarun, Nahar, Iraq, Khorasan etc. used to reach there instantly. In tussles between buyers and sellers, *Bhartiy* men, women and children, were made to move here and there like rumbling fish and fluttering birds. They were tied like animals and kept in cages. They were poked with the tip of wooden stick. The buyer would slantingly look at them, weighing their future uses, whether they would be enjoyable in fulfillment of his sexual desires or could be used like animals. Then the bargain would start. Black or white, rich or poor, small or big, this fair had only one purpose i.e. of slaves-trading.

It is pertinent to mention here that Mahmood's sword did not spare even those Buddhists who had helped Qasim and committed treason. The Buddhists were either converted to Islam or their necks were severed from the trunk with a sword. He had performed innumerable beastly, ruthlessly cruel, demonic and monstrous atrocities and orgies in Bharat. Mere imagination is enough to convulse humanity. This human-vampire suffered from tuberculosis and died in pain on 30$^{th}$ April 1030 AD and left behind a legacy of adultery, rape, robbery, arson, massacre, hellish atrocities, cow slaughter, kidnapping of children and women and their selling like animals in slave markets. Arabic historians have called these abhorrent and savagely diabolical misdeeds favorable to the *Quran* and have described him as a true Muslim. Wow! what a religion! I want to advise those buffoons, who are nourishing the crop of Islamic terror in the name of pseudo-secularism that they should, at least, know true facts. Being blinded by selfishness, they should not again push Bharat into deep well of slavery.

At this point, I want to refer to Mahmood Sayyid Salar Masood. After the death of the vampire Mahmood, Sayyid Salar Masood, known as Ghazi Miyan, along with millions of jihadi robbers, like a swarm of locust, invaded Bharat. At that time Delhi was ruled by Mahipal. Sayyid Salar Masood was very cruel and barbaric like Muhammad bin Qasim. Masood did what was directed by head Khalifa of Islam.

He slaughtered Hindus selectively, leaving only those who were circumcised or had become Muslims. Under the flag of his religious leader, this dreaded robber and murderer came to Bharat. He destroyed *Bharatvarsh* up to Meerut and Kannauj. Imagine the condition of the area which was invaded by one and a half lakh killers armed with weapons, for the purpose of looting and eliminating Hindus? His cruelty is revealed from the description made in the *Aine Masoodi* :

*Waving the flag of Bara Hazari of Satrik,*
*The caravan of Bara Hazari is moving*
*Infidels, on the way were sent to hell*
*Only those were saved who read the kalma of Bara Hazari*

The sentence "The one who read the *Kalma* of Bara Hazari" is enough to tell the horror of this aggression. Wherever this group went, that area was engulfed in chaos. Rivers of blood flowed. Temples turned into mosques. In front of their guardians, women and daughters were raped, children were beheaded and hung on spear-heads. They were driven like sheep and goats and sold in the shops.

After the death of Mahmood Ghazni and Ghazi Miyan, there was an invasion from Multan side by Muhammad Ghori in 1175 AD. He not only looted but also took concrete measures to make Bharat a permanent slave land. With the help of the traitor and deceitful Raja Jaichand of Kannauj and General Hahuli, in the year 1192, after defeating Prithviraj Chauhan, he made one of his slaves named Qutubuddin Aibak, the emperor of this place. Later, King Jaichand of Kannauj was also put to death. Historians say that Prithviraj defeated and imprisoned Muhammad Ghori several times earlier, but due to his generosity and hunger for self-praise, Prithviraj kept releasing Muhammad Ghori. This historical mistake eclipsed the glorious sun of Bharat and Bharat became a slave for thousands of years. Here I consider it appropriate to quote a part from one of my poems :

अगर मुहम्मद गोरी को हम क्षमादान न देते,
पहली बार पकड़कर उसका शीश कलम कर लेते,
पराक्रमी भारत का तब इतिहास और ही होता,
जगत्गुरु की मान प्रतिष्ठा भारत कभी न खोता।।

*Had we not forgiven Muhammad Ghori.*
*Had we beheaded him at every cost.*
*History of mighty Bharat would have been different.*
*Prestige of world Guru would not have been lost.*

According to historians, Muhammad Ghori, with the help of traitors, took Prithviraj captive. By putting handcuffs and fetters he was taken to Ghazni and his eyes were pierced. To aggravate his pain lemon juice, salt and pepper were poured into his pierced eyes. In order to exhibit his proficiency in the art of *shabad-bhedi-baan*, Prithvi Raj requested Ghori to organize a gathering. A high stage was got erected and Muhammad Ghori was asked to sit on it. Prithvi Raj, with the help of his favorite poet and general Chandabardayi, killed Muhammad Ghori with his *shabad-bhedi-baan*. After killing Ghori, Prithviraj and Chandabardayi, with their own daggers, committed suicide.

According to some historians, when Prithviraj was being tortured, Muhammad Ghori was reminded of the generosity towards him by Prithviraj Chauhan. Muhammad Ghori laughed and termed his generous behavior as foolishness. He said that he was not foolish to release the captive enemy. Ghori's response was in accordance with his character. Only by attacking Prithvi Raj's sleeping army Ghori could imprison him. In the straight field fight, Muhammad Ghori had lost every time. Unfortunately,

Prithviraj believed in morality and followed the rules of war and also believed that the enemy would do the same. According to the character of the enemy, *Bhartiys* could not change their strategies. Had we followed the saying of 'शठे–शाठ्यं समाचरेत' (be wicked with the wicked) neither would we have suffered thousands of years of slavery, nor could these vampire-invaders forcibly convert our brothers to Islam; nor could Bharat have been atrociously, dreadfully, treacherously oppressed.

After conquering Delhi, Muhammad Ghori left his slave Qutbuddin Aibak as his representative. Soon, Qutubuddin Aibak was in trouble. Being sensitive to his troubles, Muhammad Ghori again came to Bharat. He attacked Kannauj and Jaichand had to taste the result of his betrayal of the *Rashtr* and was killed. Muhammad Ghori began to create a ruckus and commotion. He went to Varanasi, looted the temples and destroyed the idols. Earth turned red with blood. Wherever he went, like any other Islamic fanatic he openly wreaked havoc.

Shri P.N. Oak wrote on page 104 and 105 of his book the *Muslim Sultans in Bharat-I*, "Desperate Muhammad Ghori was now in a religious impulsive hysteria. Scattered heads could not be counted. Earth also got tired of drinking blood. Thousands of Hindu women were dragged away from their families and were looted. The minced meat of the babies was scattered all around like cut lambs.

The Muslim army looted thousands of Hindu temples and converted them into mosques. For the second time, the holy *Shivsthan* was a victim of Muslim hooliganism. Earlier, 150 years ago Ahmed Nithaltijin had looted and ruthlessly ruined it. This loot of Ghori was the second instance of Muslim loot. The Vishwnath temple of Varanasi, the royal palace of Jaichand, the citizens and merchants were looted and a huge mountain of gold and silver was raised in front of Muhammad Ghori. To celebrate the massacre and slaughter, the Muslim army, created a catastrophe in the city. There was no house without threat or fear."

After attacking many times, in his own wicked, villainous and dastardly way, along with Qutubuddin Aibak, he established a permanent Islamic power in Bharat, and made every effort for about 700 years to destroy Hinduism. In order to exterminate

Hindus, every subsequent Muslim ruler caused rivers of blood to flow.

Some historians have expressed an opinion different from the story about the death of Muhammad Ghori in Chandabardai's Prithviraj Raso. According to them, Muhammad Ghori died on his way back to Ghazni on March 15, 1266. They are of the opinion that on the above date, on the way to Ghazni, Muhammad Ghori halted at Damyak. A small group of Hindu braves, like a thunderbolt, entered in the camp of Muhammad Ghori. In one stroke they threw Ghori's head to the ground. Whatever may be the case, there is no difference between the date and fact about the end of Muhammad Ghori. Finally, a cannibal butcher's life ended. But he had handcuffed Bharat in fetters of slavery which could not be undone for many centuries.

Before proceeding further let us glance at the life of Aibak. After being sold many times in slave markets, Qutbuddin Aibak finally landed in the hands of Muhammad Ghori. Muhammad Ghori kept him with himself. Gradually Aibak was successful in winning the trust of Muhammad Ghori. According to many historians, Aibak became the favorite of Ghori as Aibak was purchased for the purpose of unnatural sex habits of Muslims. In any case, he was even more ambitious and cruel than Muhammad Ghori. He also did what the invaders had been doing i.e. killing innocent Hindus in the name of religion; raping women and minor girls; robbing children; filling the treasury with loot; forcibly converting Hindus to Islam; converting temples into mosques; destroying idols; humiliating Hindus and spraying land with innocent blood. Aibak, rather persecuted Hindus more than any of the previous invaders.

From 1206 to 1210 AD till his death, he left no stone unturned in shedding blood. He committed these massacres only to be called a 'true and religious Muslim'. Elliot and Dawson's book two on page 226 quotes from Sadruddin Hasan Nizami's the *Tajul-Maasir*, "Qutubuddin Aibak is a Muslim and a pillar of Islam. He is the destroyer of the infidels. He devoted himself to overthrow the non-believers of their religion and the state i.e. (Hindus). He drenched the land of Bharat with the blood of the hearts of its people. On the doom's day, the Islamic believers would have to cross the river of blood by boats. Any fort or stronghold he attacked, he took it in his possession. Its foundations and pillars were trampled under the feet of elephants and turned to dust. He beheaded the crowned kings and used their heads as a crown of crucifixes. With the mighty edge of his sword, he threw the whole world of idolaters into the fire of hell. In place of idols and statues, the foundations of mosques and madrasas were laid."

He imposed the burden of Jizya tax on Hindus and thereby forced them to become Muslims. For 15 years, he was the representative of Muhammad Ghori. After the death of Muhammad Ghori in 1206, he became completely independent and declared himself to be the Sultan and ruled for four years. He died a painful death while playing polo in 1210.

After the death of Qutbuddin Aibak, his son Aramshah ascended the throne. In 1211, the Subedar of Badaun, who had been bought along with Qutubuddin Aibak as a slave revolted and captured Delhi and became the Sultan. By accumulating power, he massacred Hindus in areas like Bengal, Sindh, Gujarat, Gwalior, Ujjain, Mewar etc. and destroyed temples and converted them into mosques. Pleased with this massacre, the chief Khalifa sent a dress and Khillat as a mark of respect to him. Khalifa bestowed the title of emperor on him.

On page 146 of his book the *Muslim Sultans in Bharat Part-1*, Oak has written, "The Muslim robbers who persecuted, killed and robbed the infidels (Hindus) have always been patronized by the chief Khalifa. He now realized that the slave of the slaves, Shamsud Din Illtutmish, who assumed the title of king, had become fully entitled to receive Islamic award."

He has quoted from *Minhaj-as-Siraj*, "Wearing dress from the throne of Khalifa the messenger reached the border of Nagaur, on a Monday of 1229 AD. He entered the capital and got it sanctified. The kings, their noble heroes, their sons, other nobles and servants were honoured with clothes received from the center of Islam."

It is clear from these examples that by committing inhumane atrocities, genocide and forcibly converting Hindus into Islam and turning temples into mosques, raping women and selling them as slaves, Illtutmish represented the Khalifa like his predecessor Islamic invaders and rulers. That is why the Muslim buffoons praise him. Illtutmish had destroyed many important temples like Mahakal temple of Bhilsa and Ujjain and took away their idols, which were installed on the steps of a mosque so that the believers, that is, Muslim worshipers, could attain peace by rubbing their shoes on them. Untill Allah took his life in 1236, to quench his thirst for religious fervor he continued to plan and execute plundering with the blood of innocent Hindus.

After the death of Illtutmish, the maid's son, Ruknuddin Firoz Shah ascended the throne, but his fall began six months and twenty-eight days later. While he was out of Delhi to suppress a rebellion, his half-sister Razia, gracefully ascended the throne. The sensual chieftains who wanted to enjoy Razia's youth helped her ascend the throne. Razia sent an army and took his brother Ruknuddin as a prisoner and was brought to Delhi and sent to Allah. Razia ruled for three years and six days. Her entire time is filled with murder and looting. She also saw Hindus through the traditional Muslim lens. She too committed atrocities on Hindu. With her body and personality she entertained all the sensual courtiers and many Muslim chieftains. Malik Altunia kept Razia as concubine and later murdered her.

After the death of Razia, her brother Muizuddin Bahram Shah ascended the throne. He reigned for two years only.

Muslim chieftains also imprisoned and killed him. After this, the grandson of Illtutmish became the king and ruled for five years. He was imprisoned by his relatives and Muslim chieftains parceled him to Allah. After that the youngest son of Illtutmish ascended the throne. From 1246 to 1266 AD, this Sultan, named Nasiruddin Muhammad razed Hindus to ground. With the help of his general Ulugh Khan alias Balban, he shed rivers of blood in the Hindu towns and villages of his kingdom. By massacring the infidels, he performed his religious duty and received the love of Allah. As per the *Quran*, the one who destroys the infidels from the roots or breaks the necks of the pagans is the true believer of Allah. Nasiruddin's life also ended at the hands of his own General Balban (Ulugh Khan) who poisoned him and became the Sultan himself.

From 1266 to 1287, Balban as Sultan, did what Allah had ordered for him in the *Quran*. P.N. Oak on page 206 of his book *Muslim Sultans in Bharat-1* has described the acts of Balban, "With a torch in one hand and a sword in the other, the reign of Balban from 1266 to 1287 was literally a bare-blood-dance of the devil." For twenty one years, he continued to shed rivers of blood, rape of women and kidnap of children and set their houses on fire and these after the whole city was ruined.

On page 146 of the book *The Story of the Mughal Emperors,* Shri Ayodhya Prasad is impelled to write about Balban, "Balban destroyed the Hindu Mewatis in lakhs as they were rebellious. He set their villages on fire. Only when these Hindus of Mewat became completely calm and Muslim, they left their chase."

In a book titled *A Look at the Excellence of Hindu systems,* on page 79, Shri Brahmlochan Dubey writes, "He brutally oppressed the Mewatis and the Hindus of Katihar. Soldiers were ordered to attack villages, burn houses, kill all common and special people. Thousands of men were killed, the survivors above the age of eight, women and men, were taken as slaves. The human corpses were left rotting in every village and forest. Burnie has commented on this, "the rivers of their blood flowed and the stench of the dead reached Ganga River."

Even today, if you look at the current Islamic terrorism in the light the above quote, the same sentiments still prevail; it is a strong confirmation that as long as Islam is conducted according to the *Aayats* of the *Quran*, there will be no change in the world. Peace is impossible.

To get Islamic paradise, he started killing people by burning houses in every town and village, demolishing buildings, trampling down standing crops. Due to this massacre, mutilated bodies lay rotting in all areas. Historian Burnie says, "such a terror of this horror drama sat on the hearts of the rebellious Hindus that they lost their courage forever and ever."

Overall, Balban followed the instructions given by Allah in the *Aayats* of the *Quran.* He followed instructions and made the earth red with the blood of innocent Hindus and gave proof of his being a true Muslim. He enriched the rotten and wild

Muslim rule. This demon, who committed barbaric tyranny, died in 1287. As Balban was unable to give any worthy ruler, the slave dynasty came to an end with his death.

Although, for some time the sinner Kaiqubad, who was the grandson of Balban, sat on the throne, but Shaista Khan killed fornicator Kaiqubad and took over the throne of Delhi. Alauddin Khilji, the son-in-law of Shaista Khan, whose name was Jalaluddin Firoz Shah, mercilessly killed his father-in-law, his two sons and daughters and ascended the throne. From here the rule of Khilji dynasty started.

In fact, in the eyes of every Muslim ruler, Bharat was a huge poultry farm; the Hindus were the chickens, which not only gave eggs to the Muslim kitchen, but these chicken were converted to Islam.

Like other Muslim sultans, it was Jalaluddin's intention to do what was directed in the *Aayats* of *Quran* i.e. the annihilation of Hindus.

After killing Jalaluddin Khilji by deceit, his son-in-law and nephew, the human-vampire Alauddin Khilji sat on the throne. Alauddin Khilji broke all the records of barbarism and brutality. Although, all the earlier Muslim invaders and rulers persecuted Hindus, shed their blood and forcibly converted them into Islam, Alauddin left no stone unturned to excel them. It is expedient to mention a conversation between him and one of his religious advisors i.e. the Kaazi, which is mentioned in the *Tareekh-e-Firoz Shahi*.

Sultan asked the Kaazi, "What is the law for Hindus? Should a Hindu be a tax payer or tax receiver?" The Kaazi replied, "He is said to be a tax payer. If the tax collection officer asks for silver, tax payer should give gold to him with humbleness, great respect, honor and without asking any question? If the officer wants to throw dust or spits on his face, without hesitation he should open mouth. This filth is to be eaten by him. This acceptance of inferiority is expected from the *Muqadams* (tax payers). It is our duty to increase the glory of Islam... Allah hates these people (i.e. the infidel Hindus) because Allah orders to crush them. We have a special religious duty to suppress the Hindu people, because they are the staunch enemies of the Prophet. The Prophet has ordered us to kill these people, plunder and take them captive, convert these people to Islam, make them Halal, make them slaves and destroy their wealth. That great preacher, Hanif whose ideology we follow, has approved the imposition of Jizya tax on Hindus. The preachers of other schools have considered only one option-death or Islam." The Sultan said, "Oh! Kazi, you are a great scholar. It is absolutely according to the law that Hindus should be crushed and suppressed ........ Hindus will not obey or surrender, till they are not made absolutely poor. That is why, I have issued this order that every year only that much of grain, milk and curd should be left with them which is necessary for their sustenance, so that they can never accumulate wealth or get organized." (page 185 volume-3, Elliot and Dawson).

The above example delineates Alauddin's barbarism, bigotry and attitude towards Hindus as directed by the *Quran* and Islam. I wish those foolish, sly or so called intellectuals should read this carefully and get their mental bankruptcy cured before talking about *Ganga-Jamuni-Tehjeeb.*

In fact, he fixed the revenue tax up to fifty percent of the produce. Tax was fixed even on the grazing of animals. Jizya tax was charged for being a Hindu. Jizya tax was according to the number of members of the family. If we calculate at today's price, a Jizya tax payable by a family of ten persons, comes to eleven thousand rupees for poor family, twenty two thousand rupees for middle class people and fourty four thousand rupees for the rich.

Even today payment of 50% of tax is torturous and inhuman for a labourer, payment of 50% of tax was not justifiable at that time too. But the sole purpose was that Hindus should accept Islam or their right to live will be snatched. After leaving Hinduism and by accepting Islam, one would get freedom from these taxes and inhuman tortures. After accepting Islam as his religion, that person would get the right to live. Hindus were allowed to survive by paying Jizya tax by Khalifa Umar, but they were also imposed with hellish twenty conditions, some of which were :

1. New temples will not be built.

2. Will not repair old temples.
3. Will not restore the destroyed temples.
4. Will not wear ring.
5. Will not ride a horse with saddle and bridle.
6. If any meeting or *Panchayat* is being held, then Muslims will not be denied entry in it and everyone must listen to the words of the Muslim.
7. If someone dies in the house, no one should cry loudly.
8. Festivals will not be celebrated in public.
9. Hindus will not wear turban.
10. Paan will not be eaten.
11. Will not wear thin clothes.
12. If a Muslim wants to stay in the house of a Hindu, then a healthy Muslim will have to be kept by Hindus in his house for one day and one night, an unhealthy Muslim for three days and three nights. Hindu daughter-in-law will not complain of adultery or daughter will also not complain of rape. These crimes will not be considered as crimes in any way.

Accordingly orders were passed by Alauddin Khilji. Imagine that a Muslim could enter a Hindu's house and do anything. No one could even complain. The condition of a Hindu home was worse than that of a brothel. A prostitute is in any way prepared for it. The Hindus had to live such a disgusting life to survive.

Finally, in 1316, Alauddin, the most brutal fanatic demon of the Muslim chain, died. But he carved the story of animalism throughout Bharat. In proving himself to be follower of the *Aayats, he bathed himself and the earth with the blood of Hindus. Throughout his life, he endulged in destroying temples, hurting cows, breaking idols, plundering women's clothes and dastardly exploits. In this way he continued to worship Islam.*

*After the death of Alauddin, his powerful general Malik Kafur declared the child prince Shahabuddin as the Sultan and declared himself as the protector. After some time Malik Kafur was assassinated. After that, child Sultan's brother Mubarak Khan got him killed by taking out his eyes and ascended to the throne. Qutubuddin murdered Mubarak Khan and became the*

*Sultan. Qutubuddin had a taste to dress like women. He used to enjoy unnatural sex and had abducted a handsome Hindu boy, forcibly converted him to Islam, and kept that boy for anal enjoyment. He had great trust in him, but this boy was deeply hurt by the atrocities being inflicted on him that he resolved to kill the Sultan. He fulfilled his resolve. He killed Sultan and ascended to the throne. His name was Nasiruddin. After declaring himself a Hindu, he renovated Hindu temples in the palace, installed idols, banned the killing of cows, gave many orders for the protection and honor of Hindus, but just within two months Muslims uprooted him. The ray of Hindu hope that had shown, died away within two months. An Islamic fanatic named Ghiyasuddin Tuglaq captured Delhi and started writing the script of Islamic animalism again.*

*Ghiyasuddin Tuglaq was a violent animal like other Muslim rulers. He made the land tax more stringent. Hindus were subjected to extreme torture. In order to skin the poor Hindu masses he imposed many new taxes and made arrangements to collect them ruthlessly. The unfortunate Hindus had to sell their belongings. He shed so much blood of Hindus that Amir Khusrau was impelled to write, "Its land has been purified and cleaned with the water of the sword and the clouds of infidelity have been removed from here."*

*The clearing of the clouds of infidelity simply means that lakhs of innocent Hindus were slaughtered and thrown away in the name of religion. It also meant that Hindus should be completely eliminated. For this purpose, Ghiyasuddin Tuglaq also committed in mass massacres.*

# Tyranny of Mughals

Mughal rulers, who were followers of Islam, were not any different from their predecessors. There was a competition amongst them to set an example of cruelty. To annihilate Hindus and to prove their loyalty to the Khalifa, every Islamic ruler made similar efforts. For this purpose, shedding blood of innocent Hindus was a common modus operandi. In 1526 AD, the last ruler of the Lodhi dynasty, Ibrahim Lodhi, was eliminated by Babur, who was the first ruler of the Mughal dynasty, which ruled for 332 years. In this dynasty, people of the same family captured power by revolting against their own ruling elders, by way of either murdering or imprisoning them. This tradition of revolt against the father by the son or against uncle by the nephew continued throughout the Mughal period. Humayun captured power by looting all the property of his father Babur. Humayun's son Akbar was thirteen years old when Humayun died. Humayun continued to be a victim of rebellion by his brothers during his reign. Jahangir also unsuccessfully tried to poison his father Akbar. He revolted afterwards. Jahangir's son Shah Jahan also did what his father had done against his father. Shah Jahan's son Aurangzeb imprisoned his father and put his brothers to death.

Under Mughal rule, Hindus got the same treatment which they had during the period of other Islamic rulers. The blood of the innocent was shed and property, women, girls, boys etc. were looted as the *Mal-E-Ganimat* and they were treated in the same way as the previous rulers had done.

*Bhartiy* pseudo-historians take pride in proclaiming Akbar as the great ruler whereas the facts are just the opposite. Akbar, in a very diplomatic way degraded and debased Hindus and continuously promoted Islam in every possible way. With regard to Akabr's political debauchery, Smith, on page 160 of his book said that the whole plan was the result of the development of debauchery, pretentiousness and autocratic capriciousness.

The Christian priest present in Akbar's court has also mentioned Akbar's filthy act of asking general public to drink his feet-washed-water. Smith, referring to Pastor Xavier on page 189 of his book, has said, "Xavier has written that Akbar presented himself as a prophet. For this, the public had to believe that patients get cured by drinking his feet-washed-water. Smith has referred on the same page to the then documentary writer Badayuni, according to whom this particular type of abusive behavior was meant only for Hindus. Badayuni says, "If people other than Hindus came and expressed their devotion, Akbar used to chide them." It is clear from this quote that even during Akbar's time the plight of Hindus was equally bad and a well-planned work of destroying them was being undertaken by Akbar.

Famous historian Purushottam Nagesh Oak, in Part-2 of his book the *Muslim Sultan in Bharat*, on page 104, has accurately described this attitude of Akbar : "Akbar's marriage with many royal Hindu women has been presented in a distorted way as a grand example of his spirit of cooperation and tolerance. Rather it was like sprinkling of salt on a wound. Akbar equally indulged in sexual orgy and debauchery." He further said, "Akbar considered his entire kingdom to be a very big *Harem*. He would forcibly take ladies from royal houses of the defeated kings. This was one of his many ways to express hatered for his victims. Forcing Hindu women into their *Harems* has been the detestable practice of all the invaders.

Smith, on page 250 of his book, describes Akbar is horrifies : "The types of capital punishment were- crucifixion, trampling under the feet of elephants, beheading. As a form of minor punishment amputations and terrible whippings were normally ordered." During Akbar's tenure no Muslim girl was married to any Hindu. This is a proof that even Akbar was mentally no less a fanatic than his predecessor Islamic rulers. He himself burnt the religious buildings of Banaras and Allahabad. It is recorded in history books. Therefore, even during the reign of Akbar, Hindus were tormented and destabilized.

Due to these reasons the identity of the Hindu was weakened and vices started gripping the *Bhartiy* society.

With their pseudo-liberal policy, the Mughal rule, especially during the time of Akbar, which some historians conspiratorially describe as 'great', had completely destroyed the *Kshtriys* of Bharat. They left no stone unturned in destroying and maligning *Bhartiy* self-esteem. Islam kept constantly surging in Bharat. The radiance of the son's of Bharat became pallid. Summing up cultural degeneration in a few line is not possible. Deliberate attempts to make Akbar great by destroying the glorious past of Bharat is disgusting. The famous historian Vincent Smith called this effort "emotional futility". Vincent Smith has also mentioned on the 17th page of his book that Akbar was a foreigner in Bharat. There was not a single drop of *Bhartiy* blood in his veins. Describing Akbar's personality on the same page of his book, Vincent Smith says, "Akbar's sycophants have not given any description of his alcoholic tendencies."

After Akbar, atrocities by Aurangzeb reached the climax. Shedding the blood of innocent Hindus and converting them to Islam was his spontaneous nature. Aurangzeb also did the work of destroying many temples. He got Hindu kings brutally killed through conspiracy. Crude savagery committed by Aurangzeb in the name of Islam knew no limits.

Muslim writer Saki Mustaed Khan has written in his book the *Misar-e-Alamgiri*, "On 18 April 1666, Aurangzeb heard a rumor that *Brahmans* of Thatta, Multan and Banaras had the habit of explaining books like the *Veds*, *Upanishads*, *Bhagwad Gita* and other Hindu epics and Muslims used to go there from far off places. Therefore, he ordered to destroy the Hindu temples and schools. According to that order, the Vishwanath temple of Banaras was destroyed."

In fact, destroying Hindu temples and converting them into mosques had been done instinctively by all Islamic invaders.

I have described such cruel and heinous acts of Aurangzeb in the chapter titled "References from Sikhism". Hence, repetition can be omitted.

The Islamic rulers not only sabotaged the *Karm*-based *Varn System* of the Hindus with their nefarious practices, but they also pushed the Hindu society in a deep well of utter distortions by dividing it into sub-castes and creating untouchability through royal decrees.

As a result, for a long time, Bharat remained victim of Islamic bigotry. Inhumane, scathing, ghoulish, caustic and piercing atrocities committed by all these invaders are screaming evidences that expose how the Muslim rulers tried to completely crush and mutilate Bharat's sacred culture, philosophy and social harmony. It is amazing that even after suffering the most heart-wrenching brutality for such a long period, the Hindus are alive and their culture has survived. It is extremely astonishing. In fact, during this period, the work of distortions continued and our sacred beliefs were shattered. Historical records testify that the order of the Muslim rulers led to the birth of new castes. The gallant heroic castes who had valiantly resisted these Islamic invasions were through mandates barred from taking water from the public wells. Other Hindus were instructed that they would not be spared if found socializing with these castes. As a result, untouchability was born. Due to isolation and to save their lives some people fled to forests and those who stayed behind had to bear the brunt of untouchability.

Astu! Naturally, after so many brutal atrocities, the *Varn System* got shattered and a perverted system emerged. This system nourished hatred. Concept of class consciousness and untouchability dominated this period of Muslim invasions.

After the end of the last ruler of Mughal dynasty, Bahadur Shah, the British government used the distortion in the *Varn System* as a weapon and worked to divide the Hindu society permanently. Their well-known policy was 'divide and rule', and to implement this policy, they gave a permanent structure to the distortion of the *Varn System*. Unfortunately, writers, historians, critics, who call themselves intellectuals, epitomize these distortions as the real nature of Hindu *Varn System*.

## British Conspiracy

After the end of the Islamic rule, the British, through the East India Company, fraudulently established their rule in Bharat. Like Islamic rulers, the British too perpetrated injustice and atrocities. Adopting the policy of 'divide and rule', they kept using the distorted *Varn System* as a weapon and pitted our own kinsmen against each other. To completely erase the memories of the glorious past of Bharat from the minds of the new generations, they worked on a long-term plan. While on the one hand they shed the blood of the innocents, on the other hand, they equally tirelessly distorted our cultural heritage. For this purpose, they raised a regiment of sponsored writers and historians. They desecrated the scriptures by adding interpolated passages. Colonel Tod, as a political agent, came to Mewar in 1806 and carried out this campaign with great effort. He used *Bhartiy* writers as his weapons. He got the history books rewritten. He knew well that it was absolutely essential to distort the culture, literature and scriptures of any nation and get the glory of the past removed from the history books to make its subjugation permanent. At the behest of the Crown, they drove the leading men of Hindu society like princes, writers, historians, litterateurs, etc. in the sewerage of wine and women. A conspiracy was hatched to undermine our glorious past. They ridiculed Ram and Krishn, the *Ramayan* and the *Gita* and the *Mahabharat* by calling them fictional. They also tried to delete these from the pages of history.

The education system was changed radically. Max Muller, translated the *Rigved* with malafide intensions. Colonel Tod and Lord Macaulay tried to completely subjugate the *Bhartiy* pride. All of them had a fantasy to corrupt the illustrious generations of Bharat. They wanted to prepare a new generation which will look *Bhartiy* in appearance, but their mind and heart will be in sync with western culture. These would act as a bridge between the British rulers and the ruled. For this Lord Macaulay traveled all over Bharat and collected information from various sources regarding ways of life, customs, traditions, history and ancient literature. Accordingly, a plan was finalized with the intention to establish lasting servitude of Bharat. How detailed was his meticulous assessment of Bharat, can easily be estimated as this information was bound in ten thick volumes. After changing the education system and history he went back to England. He gave a speech in the British Parliament which continued for several days, detailing his purpose and work in Bharat. He took more than eighteen hours. The seriousness of this speech is evident from the fact that the then Queen Victoria of England heard Lord Macaulay's speech sitting in the state gallery. She heard the speech daily and also noted down the special points of his speech. Subsequently, this speech of Lord Macaulay became the foundation for policy making for Bharat in England.

Lord Macaulay opined, "I have instilled six points in the minds of the people of Bharat. As the time passes, each point, like an ocean, will destroy all the shores of dignity and modesty of Bharat. It will end the civilization and culture of this ancient country and make it non-existent."

The points of Lord Macaulay, which made Bharat non-existent are as follows :

1. UNCRITICAL
2. UNSCIENTIFIC
3. UNHISTORICAL
4. IRRATIONAL
5. LEGENDRY
6. MYTHOLOGY

It is clear that, through very cunning means, the crooked Britishers put a question mark on the veracity of all our *Veds*, *Purans* and *Shastrs*.

What is surprising is that even today, with utmost respect some people are nurturing poisonous tree of education system imposed by Lord Macaulay. History is witness that more than a hundred countries which became independent along with Bharat, discarded the British education system within two years and implemented that education system which was in sync with the culture and spirit of their own country. Unfortunately, Bharat could not do this. This is the reason why *Bhartiy Dharm*, culture, civilization, arts, skills, knowledge, science etc. are losing ground. At an uninterrupted pace, the poison of hallucinations is spreading in the whole society.

Sometimes a thought flashes in my mind as to what were the compulsions due to which the rulers who took over the reins of independent Bharat, instead of antidoting this poisonous tree kept it nurturing. I think, either they were engrossed in the enjoyment of power that they could not pay attention to it or they did not have enough courage to uproot this poisonous tree or they were in sync with the plan of Lord Macaulay. The *Manas-Putr* of Lord Macaulay, for the emotional exploitation of the masses imprecated Macaulay's education system but could not muster courage to replace it with an indigenous *Bhartiy* education system.

Had the people, who took over the reins of independent Bharat noticed the reaction of British Parliamentarians to Macaulay's speech, they would have seen through the agenda of Macaulay. The members responded by saying, "Our largest army with most modern weapons could not have been as victorious, what Mr. Macaulay has achieved alone. Mr. Macaulay did not only cut the tree of *Bhartiyta* with a saw, but he also shredded its roots so that it can never be green again. The Western education system will further ulcerate it. In such a situation, due to its foul smell *Bhartiys*, the protectors of this tree will themselves uproot it completely."

Whenever there is an attempt to end Macaulay's education system even after many decades of independence, innumerable people come out in support of it. When I, along with Dr. Murli Manohar Joshi, took the initiative to change Macaulay's education system with that of *Bhartiy* education system, these psudo-seculars created a stir by saying 'education is being saffronised'.

Colonel Tod and Max Muller openly tempered with *Bhartiy* contemplations, philosophy and history. Max Muller's commentary on the *Rigved* is devoid of scholastic knowledge. These commentaries are unrestrained babble two-faced purpose with a deceitful purpose. He, in his cunningness translated the

*Rigved* erroneously. The letter written by Max Müller to the Secretary of Bharat, the Duke of Argyle, clearly shows his sly mentality. He wrote, "The ancient religion of Bharat is doomed and if Christianity does not step in, whose fault will it be?"

Max Muller refrained from commenting on the *Rigved* not out of reverence but to spread confusion in the *Dharmik* sphere of Bharat. It is unfortunate that by using his commentary as a weapon, the conspirators are engaged in spreading casteism and disbelief in society. Similarly, with the same ambidextrous intensions, Colonel Tod got a few history books rewritten. These historians made every effort to underestimate the history of Bharat. In spite of all the cattiness, viciousness and vindictiveness these historians could not assess Bharat's history for less than five thousand years. This assessment process was driven by greed and conspiracy. Similarly, some historians did not consider both the *Mahabharat* and the *Ramayan* as a part of history. Now, irrefutable evidences have been found which are sufficient to prove that both the *Mahabharat* and the *Ramayan* are authentic religious and historical texts.

Through a detailed discussion of adverse effects of British rule, it can easily be understood that the British left no stone unturned to convert Bharat into Christianity. Along with a carrot and stick policy, all other methods were used by the British to spread Christianity. Our own brothers and sisters helped and are still helping them by being in league with them.

All the Muslim and Mughal invaders, British and secularist were and are equal partners of the perfectly hatched and voilently implimented policy of proselytization, destruction of *Bhartiy* culture and knowledge by distorting a scientifically developed *Varn System* of Bharat.

# PART-III

# Problems and Solutions

The task of transforming the distortions of untouchability and casteism into a storm was carried out in full swing during British rule. After independence, *Manas-Putr* of Lord Macaulay are engaged day and night in this disgusting conspiracy, which is a blot on the country and society. In relation to the *Varn System*, the sacred fundamental concept has been kept distorted. Bunker mentality and negative tropism of people is involved in it.

Nowadays, the whole world is battling with Islamic terrorism. UK, Russia, America, Israel and Bharat are dwarfed by the Islamic terrorism. Osama bin Laden, a synonym of Islamic terrorism, gained enormous power by spending billions and trillions of dollars. Even after sacrificing the lives of thousands of its citizens, powerful nations like America and Britain could not harm him for long. Though sometimes ago, America using its military abilities killed him. The followers of that evil-vampire, by becoming a replica of him, are challenging the whole world. Not a single day passes when blood of the innocents is not being shed somewhere in the world in bomb blasts.

It seems that Islamic terrorism has become incurable. Islamic terrorism is the result of the instructions given by the *Quran* against the infidels and of its expansionist policy. Without revealing and divulging the truth, the idea of peace and positive thinking is meaningless. In the hope of getting the bulk of the Muslim votes, some political parties in Bharat are nurturing Islamic terrorism. That is the reason why some politicians are making open statements every day in favor of Muslim terrorists. In fact, they have pushed the Hindu society into a skittle alley by dividing it into castes and cutting Muslims from the main stream of the nation and turning them into mere vote banks. A leader is now recognized by the acceptability of his caste. The appeasement policy, in the greed of votes, has damaged the *Rashtriy* spirit. Resultantly, terrorism is being encouraged in Bharat. Since the year 2000, there have been more than eighty terrorist attacks on Bharat in which thousands of people have been killed and thousands have been disabled. Unfortunately, situation is going from bad to worse in various parts of the country. The then *Bhartiy* government had completely failed to stop the bloody incidents of terrorism and sabotage. From the series of fearless terrorist attacks, how weak our polity had become. It was completely self-centered and got caught in the web of petty negative politics. The so called regional parties are also victims of this creepy and objectionable mind-set. Most of the political parties of Bharat have started flowing in the suicidal stream of caste politics. The poisonous tree of casteism that was sown during the period of Islamic rule, was nurtured by the British. The politicians and political parties are not only constantly watering the caste-tree but have turned into its nursery as well. The situation has deteriorated to such an extent that even in the present times, the caste-forest is as wide spread and dense as it was during the reign of invaders. Unfortunately, these politicians neither respect the dignity of Bharat nor are bothered about her future. To grab power they have blind-folded themselves and are sacrificing everything to weaken the core fabric of Bharat. Extra facilities are being provided to appease Muslims and Christians. These politicians, by hook or crook, want to weaken the native people of Bharat. By following the appeasement policy for Christians and Muslims and providing extra facilities to them, they are openly humiliating and slandering the native Hindus. By dividing society into castes and sub-castes these politicians have dug a deep gulf between people of Bharat so that their consolidated power does not affect their governmental powers. To keep their power structure intact they follow the appeasement policy for Muslims and Christians. The caste enmity which was not visible even during the time of British and Islamic rule, is visible in the form of orgy, which is a terrible threat to the future of Bharat.

Terrorists are being cultured within and outside Bharat to make the expansionist policy successful through Islamic terrorism. To convert Bharat from 'Darul Harab to Darul Islam' a lot of money is being pumped by Islamic countries. It is being provided to various organizations. The so-called secular leaders of *Bhartiy* politics have turned a blind eye to this fact.

I would like to congratulate those Muslim scholars, who do not recognize the wrong interpretations of the *Quran Majeed* and the wrong traditions of Islam. The extremists continue to issue Fatwas to kill them. Of these, Salman Rushdie, the author of *Sanetic Verses* and Taslima Nasreen, the author of the novel *Lajja*, are notable examples. Taslima Nasreen is still living in the shadow of terror. Regrettably, the international community is not coming forward with the intensity with which it should come out against this jungle law. Because of opposition by a few, this political appeasement culminated in banning the books *Satnic verses* and *Lajja*. To keep the fanatics happy and to keep their hold on the vote bank the so-called pseudo-secularists have created a hurdle in providing an asylum to their authors. These pseudo-secularists are ever ready to tarnish the faith-centers, belief-systems and convictions of Bharat. Insulting the native society of Bharat in the name of secularism. Some politicians and political parties, who have become a part of international conspiracy, are attacking Hindus to openly. Temples, other religious places, saints and Hindu organizations are their targets. This is being done to please Muslims and Christians. I am sure that the black money coming from abroad is also a part of their booty. Blinded by malignancy, they are neither able to see Bharat becoming weak, nor do they see the tears of Mother Bharati.

In the plan to turn Bharat from 'Darul Harab' to 'Darul Islam', these political parties and politicians, have also adopted various other means besides terrorism.

**Infiltration**

Islamic countries are constantly working to infiltrate Muslims into Bharat. The religious institutions of Islam, the Mullahs and clerics are engaged in carrying out this work with full force. They are blessed by certain politicians and political parties of Bharat. While lakhs of people from Pakistan have chosen the border areas including Kashmir for this, people from

Bangladesh have infiltrated into Bengal, Assam, Mizoram, Bihar, Jharkhand etc. and into the interior parts and big cities of Bharat. Their number is believed to be more than three crores. Most of them have also got their names registered in the electoral rolls of Bharat. In many areas they play a decisive role.

In connection with the infiltration from East Pakistan an article appeared in the Jhansi edition of Dainik Jagran on November 11, 2008 written by Shri Jagmohan, the country's famous thinker, a former Union Minister, Indian Administrative Service officer and the former Governor of Jammu and Kashmir. I consider it appropriate to quote from this article titled "Things Slipping Out Of Hand."

"Politicians from East Pakistan, now Bangladesh, started spreading disease of land-grabbing agenda in Assam. Sheikh Mujibur Rahman openly declared that since Assam was rich in forests, minerals, resources and petroleum, Assam should be merged with East Pakistan for economic strength. Parliament's attention was drawn to infiltration as early as 1950. The then Home Minister Vallabhbhai Patel realized the seriousness of the matter. Taking quick action, he got an Act passed in 1950. But soon after Sardar Patel's death in December 1950, the Act was repealed, because issues were raised in opposition of this act. This was the first major blow to Assam's interests and efforts to protect the nation. Security was jeopardized to such an extent that during the Chinese invasion of 1962, a large section of people hoisted the Pakistani flag in Assam. In view of such incidents, the anti-infiltration scheme was implemented. Accordingly, a tribunal was setup to identify the infiltrators on the basis of the National Register of Citizens of 1951. The then Chief Minister V.P. Chaliha implemented it with full gusto. Two lakh forty thousand infiltrators were identified from 1964 to 1967. Apart from this, between 1967 and 1970, twenty thousand and eight hundred Bangladeshies were identified. But narrow political motives came to fore. Fakhruddin Ali Ahmed propagated that if Chaliha continued this campaign, the Congress party would lose the Muslim vote not only in Assam, but also in the whole country. As usual, the vote bank politics won. The plan to stop the infiltration was virtually abandoned and the tribunal was dissolved. It was a single-handed victory for the forces involved in the infiltration.

The policy of openly supporting Bangladeshi infiltrators came to light in 1976-80 when elections were conducted on the basis of the electoral rolls of 1976. A large number of Bangladeshies were also included in the voters list. The Chief Election Commissioner himself publicly stated that the population of Assam increased by thirty five percent between 1961 and 1972. As a result, dissatisfaction prevailed among the people of Assam and a mass movement started there. The movement was started by the 'All Assam Student Union' but United Liberation Front of Assam (ULFA) resorted to the path of violence. The Legislative Assembly election of 1983 added fuel to the fire. The period between 1980 and 1984 witnessed spike in the violent incidents. These included the infamous Nellai and Gohpur massacres. Woefully, even during this bloody phase, the ruling party at the Center could not break free from its narrow political insistence on keeping the vote bank intact. On the contrary, it continued its efforts to provide legal shield to the suspected infiltrators. The Illegal Infiltration (Determination by Tribunal) Act was passed in 1983. The provisions of the new law were such that identification and expulsion of infiltrators became extremely difficult. After the Assam Accord, elections to the Legislative Assembly were held again. In which the Asom Gana Parishad (AGP), the political arm of the All Assam Student Union assumed power. It was expected that peace would return to Assam but the performance of the AGP was dismal. The biggest failure of Assam Gana Parishad government was its failure to deport illegal Bangladeshies.

The AGP Government neither pressurized the Central Government to amend or repeal the illegal Migrants (Determination by Tribunal) Act 1983 nor did it try to drive out the infiltrators by taking appropriate action as per the provisions provided under this Act. Of the estimated thirty lakh infiltrators, only five hundred were deported to Bangladesh. The conduct of the Central Government was even more condemnable. It did not show sincerity in following the Assam Accord. The infiltration of Bangladeshies continued in Assam despite the voices being raised by the media and some other forums. In the three-year period between 1964 and 1966, the number of voters in the seventeen assembly constituencies of Assam increased by more than thirty percent and in the fourty assembly constituencies by more than twenty percent, while the average growth of *Bhartiy* voters during this period was only seven percent. One of the damaging motives of the Government of Bharat was to simultaneously adopt a hard and soft approach to solve the problems related to national security. Operation Bajrang and Operation Rhino, both the military operations were halted midway."

After reading this article, it is clear that the appeasement policy of the Congress government at the Center, far from stopping Muslim infiltration, gave a fillip to it. Consequently, in present times more than three crore Muslim Bangladeshi infiltrators have become a serious threat to the law and order and the identity of Bharat.

Most of these Bangladeshi infiltrators are indulging in theft, dacoity, looting, smuggling of arms, trading in counterfeit notes, narcotic substances and prostitution from the neighborhood. Consequently, the law and order situation of Bharat is crumpling. These infiltrators are playing a major role in terrorism. Bangladeshi infiltrators have become tools in the hands of the Islamic fundamentalists, terrorist organizations and ISI (Pakistani Intelligence Agency). As a result, not only the security of the country is in complete danger but the majority society of India is also weakening and falling apart.

During the rule of Islam from the 12th century to the 18th century, Assam was attacked seventeen times, but the invaders had to face defeat. In the battle on the banks of the *Brahmputr* in 1671 AD, the great general, Lachit Barphukan, taught the demon Aurangzeb a bitter lesson so much so that he could never cast an evil eye on Assam again. Surprisingly, Assam, which had been defeating the Mughals for hundreds of years, collapsed in front of crooked British in 1826 AD. The misfortune, which started in 1886 is continuously increasing its speed till date. In the border areas where Christians form a majority, the number of Muslims has increased indiscriminately. When India was partitioned, the Muslim forces divided Bengal. In the name of Bangla language a part of Pakistan broke away from it by taking human, moral and military support from India. Thus, in 1971, Bangladesh came into existence. At that time, the rulers had made a commitment to leave Islamic recognition and make Bangladesh a secular country, but within three years the 'secular' cloak was thrown off and Bangladesh became a radical Islamic state. Bangladesh, born out of the sacrifice of thousands of our soldiers, is engrossed in digging our roots. Not only this, the plan to create a 'Greater Bangladesh' is being implemented. Accordingly, in a planned manner a large number of infiltrations are underway.

According to the Governor of Assam Mr. Ajay Singh, about six thousand Bangladeshies infilter the borders of Assam everyday. Public statements regarding these illegal infiltrations have been made by former Assam Governor Shri SK Sinha, former Union Home Minister Shri Inderjit Gupta, the then Union Minister of State for Home Shri Prakash Jaiswal, former Chief Minister of Assam, Hiteshwar Sekia, former CBI Director Joginder Singh, former Union Home Minister PM Sayeed, former Director General of Border Security Force Shri Prakash Singh, former Chief of Army Staff Shri Shankar Roy Chowdhury, former External Affairs Minister Mr. Yashwant Sinha and the Law Commission headed by Mr. P. Reddy during the National Democratic Alliance Government. If all these statements are read together, then the magnitude of the problem of infiltration becomes frightening.

The eminent writer Jaideep Sekia had said, "The followers of Islam want to form an Islamic state by joining some areas of Myanmar and North East India in Bangladesh." He has warned "if eyes are kept closed by this region, it will emerge as a new Afghanistan in the times to come."

The main objective of this infiltration is to capture the 'North-east' of Bharat by increasing their population. Eleven districts of Assam adjoining Bangladesh, six of West Bengal and six districts of Bihar have become completely Muslim-majority districts. It is a part of the plan to create Greater Bangladesh.

Bangladesh is challenging our sovereignty by collaborating with radical Islamic organizations like ISI and Al Qaeda. Instead of curbing vote bank politics communist parties, to regain their power, are openly giving all kinds of help to these infiltrators. In this race, Congress also does not want to be left behind. The race for power has blinded many political leaders of India. They do not, intentionally, see the crisis of Indian identity caused by these infiltrations.

You will be surprised that the Congress government passed an Act 1983 (IMDT) Illegal Migrants Determination by Tribunal, in the Parliament to expel Bangladeshi infiltrators and to cool down the Assam movement. They created an illusion of driving out Bangladeshi infiltrators, but its under current was purely to protect them. That is why this law was struck down by Honorable Supreme Court of India in Sarbananda Sonowal vs Union of India on 12 July 2005.

In Sarbananda Sonowal, the Honorable Supreme Court invoked Article 355 of the Indian Constitution to strike down the I.M.D.T. Act 1983 that had placed the burden upon the state in a foreigner's case. The Court thus, established a constitutional requirement that the burden would always lie on the individual to rebut the allegation that he/she was a foreigner. The Court's rationale for doing so was that it would be difficult for the State to give an exact date of entry of a foreign national who had surreptitiously crossed the Indian border and the State could not remain a mute spectator to the continuing influx of illegal migrants. Article 355 casts duty upon the Center to protect every State against external aggression and internal disturbance.

In I.M.D.T. Act, burden of proving the citizenship or otherwise rested on the accuser and the police, not on the accused and this was a major departure from the provisions of the Foreigner's Act 1946.

I.M.D.T. Act is coming to the advantage of such illegal migrants as any proceeding initiated against them almost entirely ends in their favour, enabling them to have a document, and official sanctity to the effect they are not illegal migrants.

Repealing this law, Honorable Supreme Court said, "Therefore, we have now come to the conclusion that this law is completely unconstitutional and discriminatory. Therefore it is set aside." These comments made by the supreme judicial body of Bharat are enough to make it clear that a disgusting game was being played with the sovereignty, integrity and identity of India in the greed of votes by Congress Government.

As already mentioned earlier, the concern in the context is not only infiltration from Bangladesh, but also the infiltration from many other Islamic countries like Pakistan etc. Many political parties have chosen to close their eyes to this fact and an indiscriminate infiltration continues into Bharat.

The purpose of all this is to gradually disintegrate Bharat into pieces and make her non-existent by bringing demographic changes in population. Only then their mission to turn Bharat into Islamic state will be fulfilled. With the malafide intension to weaken Bharat, the native society is being divided, as the divided society will not be able to get organized and create hurdles in their race for power.

**Flood of Mosques and Madrasas**

From the 12th century to the 18th century a network of mosques was laid by Muslim rulers and many facilities were given to Madrasas. This was curbed after the arrival of the British. But in independent India there has been a flood of mosques and Madrasas. All over Bharat including border areas, Mosques and Madrasas are being built under a well-planned conspiracy. Intelligence agencies including Central Bureau of Investigation, Intelligence Bureau report that most of these mosques and Madrasas are centers of terrorism and hatred and are managing an undeclared war against India. Instead of halting this, the governments have not only turned a blind eye to it, but they are giving grants worth many millions to Madrasas just for the sake of votes. Such an act should be called suicidal. Vote bank politics is openly playing with the sovereignty and integrity of India. That is why the Kashmir Valley became devoid of Hindus. After their properties, businesses, houses, shops were looted, Hindus were forced to flee. These Hindus are homeless and are still struggling for livelihood in Delhi and other parts of the country. The unfortunate thing is that the Central

Government, the so called political parties and leaders, who boast of secularism, do not even talk about the return of these Hindus to their homes in Kashmir Valley. In fact, due to this, Bharat is weakening and its native Hindu society is disintegrating.

**Family Planning and Islam**

Increasing population is definitely a threat to the world. That is why the whole world vigorously advocates family planning. Many countries have achieved unprecedented success in family planning but India is lagging behind. The majority of Hindus have adopted family planning but Muslims have completely rejected it as it is considered against Islam. Consequently, the number of Hindus in India is decreasing and the number of Muslims is increasing rapidly. In the census of 1951, the number of Muslims was two crore and Hindus was forty five crores. There has been a quantitative increase in the number of Muslims in the census of 1961. According to the census, the number of Muslims is said to be more than sixteen crore. If the infiltrators from Bangladesh and Pakistan were counted in this census the number would cross twenty crore mark. The number of Hindus increased from forty five crores to only sixty five crores. The number of Muslims has increased ten times, while the number of Hindus in total is more than one and a half times. Earlier it was normal for Hindus to have seven/eight children in a family, now the scenario has changed completely. Most of the educated Hindus have accepted the norms of 'we two, our two'. On the contrary, owing to the religious belief of having four wives, the population of Muslims is increasing exponentially. The ultra orthodox in Bharat rejected family planning outright and termed it as anti-Islam. The thinking behind this is to make India from 'Darul Harab' to 'Darul Islam'. Muslims have accepted family planning in many other countries but not in India. If this momentum continues, then within 20/25 years Muslims will be in majority in India. If the majority Hindus become minority, their plight can well be envisaged. The pitiable condition of Hindus, is clearly visible where they have been reduced to minority status and Muslims are in majority. The valley of Kashmir, Assam, Kerala etc., bear witness to it. It is ardous for Hindus to live here.

Just a few days back, I was on a tour in a village of Jhansi district. A Muslim gentleman also attended my meeting. I asked him, "In this village, your family is the only Muslim family, is there any kind of problem to you?" He laughed and said that he and his family do not have any problem with the villagers. I have everyone's support. I asked the same gentleman, "Imagine the opposite. if there were Muslims and only one Hindu family in this village, would they live comfortably with you?" That Muslim gentleman himself admitted that Hindu family would not be able to live comfortably. This example is enough to understand the reality of present and times to come.

At the time of partition in 1947, two crores Hindus lived in Pakistan and there were many temples. Both, the Hindus and the temples are on the verge of extinction. The number of Hindus has reduced to a few thousand and the number of temples to a few dozen, while there has been a quantitative increase in both Muslims and mosques in Bharat. A hugely difference between *Bhartiy* and Pakistani mentality is clear perceptible. It is incontrovertible that *Bhartiy* mentality is liberal and Pakistani mind-set is ultra orthodox.

The fanaticism of the *Quran* can be found in many other sects too. Since the Christian sect originated from Jesus Christ much before the origin of Islam the Jewish sect was also dominant at that time. A fierce hatred is writ large in the *Aayats* of the *Quran* towards the pagan as well as the other sects.

At the time of Islam's emergence, there was a tribal culture, in which the rules of hand for hand, eye for eye, head for head, tooth for tooth were applicable. The general public will also agree that with the intellectual development of a person, there comes a qualitative change in the thinking of that person. If we chronologically look at development of a person's thinking from childhood, his level of thinking in the middle age will look completely different from what it was there during the childhood. Naturally, the under developed thinking of a child cannot be equated with developed and mature thinking. If we are honest and consider the phased development of our thinking, then the childhood thinking will appear to be very trifle and outdated. The same is the case with the *Quran*.

*Quran* may have come to light and got acceptance as a religion in the circumstances of those times. But today its *Aayats*, filled with hatred and violence, cannot be considered ideal. Due to scientific and technological research the world has become very small. The era of swords, spears and knives is over. Therefore, the cloak of orthodox thinking will have to be thrown off. When the whole world has started calling most of the followers of Islam as terrorists, then the Muslims have to come forrward and take a positive initiative to protest against terrorism. The international community will have to take such concrete initiatives as would eradicate radicalism from the minds of the followers of this sect. At the same time, people who call themselves intellectuals and thinkers also have to openly put up the true picture in front of the society so that evil forces do not get any support or shelter. Only the followers of Islam such as Salman Rushdie and Taslima Nasreen could dare to speak out against bigotry. But they did not get enough support from the global community. Thinkers like them, if properly encouraged and supported, can dissuade others from the orthodox and jungle law ideology of Islam. As this is not happening, other Muslim scholars, with clear understanding and positive thoughts in this context get demoralized. It is bad omen for future.

Astu! The prevelant *Varn System*, is very distorted version of the original. The major reason of distortions is the sickening anamorphic, gnarled, perverted and misleading attitude of many followers of Islam, Christianity and so-called pseudo-secularists. How a society which was systematically crushed for thousands of years, with a well-planned conspiracies to break it can remain untouched by distortions? The present *Varn System* of Hindu society is the result of this malignant conspiracy. From practical and management point of view the original concept is proper and scientific.

**Relevance of Varn System**

The relevance of the *Varn System*, in today's perspective is as apropos as it was in the ancient times. Sant Vinobha Bhave says, "Gandhi ji does not support *Varn Ashram* system just because he is a Hindu, but supports it on the basis of the economic and social philosophy. It can be said that in any society, a person has to follow some kind of *Varn*-like system. While accepting the *Varn System*, Mahatma Gandhi has always opposed class discrimination and untouchability. Dr. Ambedkar's literature also appears to endorse this fact in a variety of ways." The views of Dr. Ambedkar in his book *The Untouchables, Who were they? Why they became untouchables?* are given below:

1. Just as the difference of race is not the basis of untouchability, similarly the difference of profession is also not the basis of untouchability.
2. Though, the impure people, as a class were born, during the time of Dharmshastras, the untouchables came into existence much after four hundred AD.
3. In trying to trace the origin of the untouchability, do not mix the untouchable and the impure. A number of orthodox writers equated untouchables with impurity. It is a mistake. Untouchables and impurity are different. (introductory section)

Dr. Ambedkar declares on page 164 of this book, "If we take an account of the information we have received, it is clear that we cannot say that they were untouchables in the modern sense of the word 'untouchable', that is, what we consider as untouchable today. There is nothing about them in our religious scriptures."

Dr. Ambedkar says, "There was no untouchability in the *Veds*. As far as the time of the *Dharmsutrs* is concerned, we have seen that there was impurity but not untouchability. Untouchability is not known till 400 AD. In 600 AD there was a dynasty of *Chandals*. It is a far cry from considering them as fallen."

Dr. Ambedkar's views find full support in the light of historical evidence. It is clear that most of the *Shloks* of the *Manusmriti* are sublime and ideal, but some *Shlok* were interpolated later. The *Manusmriti* declares in the first chapter itself :

जन्मना जायते शूद्रः संस्कारेण द्विज उच्यते
विद्यया याति विप्रत्वं त्रिभिः श्रोत्रिय कथ्यते
एवमाचारतो दृष्ट्वा धर्मस्य मुनयो गतिम्
सर्वस्य तपसो मूलम् आचार। जगृहुः परम्।। 1,111

*Mahatmas have come to know from their own experience that everyone is born as Shudr but becomes Brahman through his attributes and Karm. The pace of dharm is through conduct. Without ethics, dharm cannot remain stable and cannot survive. Therefore, the Maharishis have considered attributes and Karm as the basis of all kinds of austerities.*

एवं यः सर्वभूतेषु पश्यत्यात्मानमात्मना।
सः सर्वसमतामेत्य ब्रह्माभ्येति परंपदय्।। 12,127

*The one who sees all living beings, souls and the Supreme Soul as one becomes a seer, attains the Supreme Brahm, that is, the Supreme Soul.*

अव्रतानाममन्त्राणां जातिमात्रोपजीविनाम्।
सहस्त्रशः समेताना परिषत्वं न विद्यते।। 12,115

*Thousands of people who consider themselves superior only by caste are misguided, corrupt and do not know anything about Mantr, one must not consider the views expressed by them in one voice, as Dharm.*

सेनापत्यं च राज्यं च दण्डनेतृव्यमेवच।
सर्वलोकाधिपत्यं च देदशास्त्रौविदर्हति।। 12,101

*The person who is well versed in the Veds is the right person to be the commander, king, judicial officer and all other posts of government.*

वेदशास्त्रार्थतत्वज्ञो यत्रतत्रश्रमे वसन्।
इहैव लोके तिष्ठन् ब्रह्मभूयाय कल्पते।। 12,103

*A person who knows the essence of the Veds, in whatever ashram or Varn he is, attains to Brahmantav in this world.*

From the above examples, it is clear that the *Shlok* numbers 266, 270, 271, 278, 276, 280, 281, 282, 283 added in the 8th chapter of the *Manusmriti* were interpolated later. Many more *Shloks* are in contradiction of the basic spirit of the *Manusmriti*. Because of the above-mentioned interpolated *Shloks* the *Manusmriti* became controversial. The basic spirit of the *Manusmriti* is just the opposite of these interpolated *Shloks*. The basic sentiment of the *Manusmriti* is that a *Varn* is attained on the basis of attributes and *Karm*; it is called the second birth of that person. The interpolated *Shloks* in the eighth chapter have gone far ahead in hatred than the *Aayats* of the *Quran*. It is clear that these *Shloks* were added later with the intention of defaming Hindu *Dharmik* scriptures. For petty political gains, some political leaders also use the later interpolated *Shloks* as their weapons. Just as butter is not visible in milk, but through churning, the 'essence' gets separated, similarly, when the *Manusmriti* is churned with a pure mind, the interpolated *Shloks* start separating as whey from the original. As many organizations are trying to defame Hinduism with full force and are making unwarranted allegations against Hindu *Dharm* these days, similarly, during the time of slavery, there was a systematic attempt to defile the scriptures. The interpolated *Shloks* are the end product of such defilement.

The strong evidences in the history in favour of the original *Varn System*, are sufficient to prove that the *Varn System* is the ideal system for social management. For an example, the Gupta period is considered the golden age of history. The *Varn System* was fully implemented during this period. At that time Hinduism was known as *Brahman Dharm*. The term Hinduism had not emerged at that time. The Gupta period is believed to be of about 270 years. Everyone must consider this fact that Bharat was at its peak when the *Varn System* was in full force; our downfall began when the *Varn System* slid down or we abandoned it or it was distorted. Therefore, the *Varn System* is still as necessary and indespensible today as it was at the time of its emergence.

As I have already said, while explaining *Varn Dharm*, one becomes proficient and gains expertise in any field after a lot of hard work. This expertise decides one's *Varn*.

Different types of qualities and attributes are required for different types of works and one must have proficiency in that work. Every person cannot have the same caliber and every job or work is not same. It is not possible for everyone to do everything. Using his intellectual ability and by being immersed in his study, a person can write literature or a scientist can do scientific research. If per chance they are recruited in army and get engaged in physical exercises prescribed for soldiers, will they be able to achieve proficiency? Can a soldier become capable of doing scientific research?

Qualifications are fixed for all the posts : doctor, engineer, Indian Administrative Services, Indian Police Services or any other state level services. A person has to prove his suitability through competition for his desired post. If one expects a less educated sweeper or an agricultural laborer working in the fields to pass any competitive examination, it may not be possible for them. Minimum qualification is prescribed for appearing in these competitive examinations. Present government system has also set the qualifications to get the job in first, second, third and fourth category. A manager and peon cannot be put in the same class. Nothing can be more ridiculous than this.

Four statues are placed together on a main square in China. The idols are of worker, soldier, doctor and farmer. Some democrats planned to remove these statues but government did not allow. With a change in the order of these idols, four *Varns* of *Bhartiy* system will be visible. There is no fault in the *Varn System* set by our sages. This system lacked rivalry and was based on the training or education by experts. This was not based on birth but was based on attributes and *Karm*. Wherein lies the fault???

As said in the preceding pages, Indian sages considered society to be the great form of Bhagwan. That is why feeling for everyone's welfare permeates it. 'सर्वभूत हितार्थ' (for everyone's welfare). He should engage in the well-being of all beings, animals, birds, plants. This is the lofty ideal of our contemplations. The *Bhartiy Varn System* is inspired by this basic thought.

In fact, hope and enthusiasm can elevate society, frustration and despair can cause regression. All of us should work for the promotion of hope and enthusiasm.

Unfortunately, spreading frustration, indolence and desperation is the goal of so-called intellectuals. Instead of unveiling the truth before everyone, they present the distorted version.

If we look seriously at *Sanatan Dharm* and its social structure, we see a divine pragmatic, logical, rational and beautiful *Varn System* which is not to be found anywhere else in the world. The *Bhartiy* social philosophy appears to be based on 'work as per capacity, price as per requirement.' In this self-driven system, there are building blocks for classless, exploitation-less society. It prepares the basis of suitable living for each one. Equality and harmony are inherent in our social imagination. Social harmony and justice are the life-line of this system. In Western thought, profit motivates the individual but the Indian thought and system is based on the motivation of self-realization and attainment of salvation. In fact our social *Varn System* is born out of a grand feeling and is based on attributes and *Karm*. *Karm* based *Varn System* is a classical fundamental concept which is as relevant today as it was in ancient times. There is a need to acquaint the public with the original concept of the *Varn System* of Bharat. This will end the process of social fragmentation. The conspirators will get a tight slap on the cheeks. Bharat will be able to attain the glorious place of Jagadguru of the world by bathing in the ocean of social harmony.

All men of honour, who dream of 'परम् वैभवन्नेतुमेतत् स्वराष्ट्रम्' (taking *Rashtr* to supreme splendor and magnificent majesty) should start a systematic campaign to make our *Rashtr* glorious. One day Bharat will be the leader of the world again and the whole world will bathe in her aura. Only then, the slogan of 'सर्वे भवन्तु सुखिनः सर्वे सन्तु निरामया' (all be happy, healthy) will prove to be meaningful. This storm of distortions will automatically end and Bharat will regain the glorious position of being a world leader.

# Glossary

## A

- Aatma - soul
- Aatmshlagha - self-praise
- Aavarn - covering, obstruction, veil
- Abhichar - process or technique
- Abhisar - convergence
- Adharm - inequity / immorality/ injustice
- Adipurush - is the Ishwar, also referred as Bhagwan Shiv.
- Agnihotr - offering of ghee into the sacred fire as per strict rites. A Brahman who performs Agnihotr ritual is called Agnihotri.
- Alankar - means figurative way of saying something.
- Amritputr - son of divine nectar.
- Annapind - are balls of cooked rice mixed with ghee and black sesame seeds offered to ancestors during Hindu funeral rites and ancestor worship (Shradh)
- Anusheelan - perusal / in pursuit of
- Anusthan - ceremony, ritual, function, celebration
- Ashram - hermitage
- Astu - let it be, so it be, may it be so
- Asur- demon, uncivilised
- Avarn - communities who do not belong any of the four Varns. They are also known as Pancham Varn.
- Avdhnam - observation.

## B

- Bhakt - a devout believer in something or some Bhagwan.
- Brahm - the ultimate reality.
- Brahmantav - Brahmanism, a state of being in Brahm.

## C

- Chandal - someone who deals with disposal of corpses
- Chhand - Sanskrit prosody or refers to one of the six Vedangs, or limbs of Vedic studies. This field of study was central to the composition of the Veds, the scriptural canons of Hinduism, so central that some later Hindu and Buddhist texts refer to the Veds as Chands.
- Crore - equal to 10 million

## D

- Daityas - are a clan of demons Danavs.
- Dalit - is basically a caste defined in Constitution under Article 341, listed as the Scheduled Castes.
- Dand - penalty. It also signifies staff held in hand signifying authority or renunciation.
- Dandkarany - is the name of a forest mentioned in the ancient Bhartiy scripture the Ramayan.
- Darul Islam - the house or abode of Islam. It signifies Islamic territory in juridical discussions. For the majority, it is thus suggestive of a

geopolitical unit, in which Islam is established as the religion of the state.

- Darul-Harab - is a term classically referring to those countries which do not have a treaty of non-aggression or peace with Muslims.
- Devyan- vehicle of a Gods.
- Devlok - abode of Gods.
- Dharm - an individual's duty fulfilled by observance of basic principles of cosmic or individual existence, divine law, conformity to one's duty and nature.
- Dharmnirpeksh - secular
- Dharmataran - proselytization
- Dharmcharan - religious conduct, acting according to the tenets of religion, virtue.
- Dharmik - religious
- Dwaparyug - or the bronze age, lasts 864,000 years and the process of self-realization is the worhship of the deities within temples. Godly qualities are reduced to 50% by now.
- Dwij - someone who is born twice. One is biological birth and the other is through knowledge.

**F**

- Fatwas- in Islam, a formal ruling or interpretation on a point of Islamic law given by a recognized authority/ qualified legal scholar (known as a mufti).
- Four Yugs - each Yug is an age with specific characteristics in which incarnations of Krishn appear. The four Yugs make up a cycle called Divy-Yug, which lasts 4,320,000 years. One thousand of these Yugs equal one day of Brahm, which is called a Kalp. Brahm's lifespan is 100 years of this time. In each Yug, there is a specific process of self-realization (Yug Dharm)

**G**

- Gandharvas - is a class of celestial beings whose males are divine singers and females are divine dancers.
- Gun-Dharm - something which is inherent quality or character.

**H**

- Harijan - children of God.
- Hawan - is a Sanskrit word which refers to any ritual in which offerings into a consecrated fire is the primary action.
- Hvya - offerings in sacred Yagy-fire.

**J**

- Jajia Tax - historically, a tax paid by non-Muslim populations to their Muslim rulers.

**K**

- Kahars- are a community of palanquin bearers.
- Kaliyug - the iron age of hypocrisy and quarrel lasts 432,000 years. Bhagwan Krishn appeared in His original, transcendental form right before the beginning of Kali Yug. The process of self-realization is Sankirtan, the chanting of the holy names of the Lord. God consciousness is reduced to 25% of the population and life expectancy is only 100 years. By now already 5000 years of Kali Yug

have passed and it is predicted that by the end of the Yug people will hardly be older than 20 years and their only food will be meat.

- Kand - part of a book.
- Karm - refers to both the executed 'deed, work, action, act' and the 'object, intent'. A good action creates good karm, as does good intent. A bad action creates bad karm, as does bad intent.
- Katha - is an Indian style of religious storytelling, performances of which are a ritual in Hinduism.
- Kshtriy - a member of the second of the four Varns. The traditional function of the Kshtriys is to protect society by fighting in war time and governing in peace time.
- Kumbhkonam - roughly translated in English as the "Pot's Corner", is believed to be an allusion to the mythical pot. Brahm that contained the seed of all living beings on earth.
- Kvya - offerings to ancestors.

## M

- Mal-e-Ganimat - valuables taken by violence especially in war.
- Manas-Putr - psychic-sons. Manas-Putr is a Sanskrit term which literally means willingly accepted as own son.
- Manishi - wise, clever, intelligent, learned, prudent
- Mantrdrsht - seer of Mantrs
- Manvantar - literally meaning the duration of a Manu, or his life span. Each Manvantar is created and ruled by a specific Manu, who in turn is created by Brahm, the Creator himself.
- Maryadit - dignified
- Mat - doctrine
- Maulik - basic, fundamental, original
- Mazhab - equivalent to religion.
- Mekhla- waist girdle.
- Muni - one who has understood the nature of the silence.

## N

- Nishkaam Bhaav - selfless spirit

## P

- Panchyagy - also known as Panch Maha Yagy, these are: Brahmyagy, Devyagy (worship of Gods), Pitriyagy (worship of ancestors), Bhootyagy (worship of other beings), Manushyayagy (worship of fellow humans)
- Panchamvarn - an other varn besides four varns.
- Pandit - person who has gained a lot of knowledge
- Panth - cult
- Paramhans - a Sanyasi of the highest level of spiritual development in which union with ultimate reality is attained.
- Paramatma - divine-Self.
- Parbrahm Parmeshwar - The Absolute, The Eternal
- Pitri-Tarpan - Pitri Tarpan is a way of gratifying the ancestors and freeing them from any unfulfilled desires so that they can complete their journey to the heavenly abode with peace, satisfaction and happiness.
- Pitriyan - vehicle of ancestors
- Pitrlok - abode of ancestors

- Prayshchit - is a Dharm-related term and refers to voluntarily accepting one's errors and misdeeds. It is a means of penance to undo or reduce the karmic consequences.
- Punarjanm - rebirth
- Purankar - writers or explainators of Purans.
- Purush - male
- Purusharth - an object of human pursuit. It is a key concept in Hinduism, and refers to the four proper goals or aims of a human life. Purusharth is a composite Sanskrit word from Purush and Arth. Purush means "human being", "soul" as well as "universal principle and soul of the universe".
- Purusharth Chatushtay - denotes four goals of life, righteousness (Dharm), wealth (Arth), desires (Kaam) and salvation (Moskh).

**R**

- Rishi - an accomplished and enlightened person, who after intense meditation (tapas) realized the supreme truth and eternal knowledge. They were scientists.
- Rishikas - female Rishis.

**S**

- Saamgan - to sing together.
- Samidha - an offering for a sacred fire during Hawan.
- Sampardaya - sect, 'spiritual lineage'. It relates to a succession of masters and disciples.
- Samta - parity, affinity
- Sandhyopasana - worship at the junctions of time of evening and night, night and morning.
- Sarg - canto
- Sarvhut - an offering in entirety.
- Sati - chaste woman
- Satyug - sometimes also called Karta-Yug: the golden age lasts 1,728,000 years. The process of self-realization in this Yug is meditation on Vishnu. During this Yug the majority of the population is situated in the mode of goodness and the average life span at the beginning of the Yug is 100,000 years.
- Satyam - that which is true
- Savarn - belonging to four Varn.
- Shabad-bhedi-baan - word piercing arrow
- Shaivism - those that revere Shiv as supreme are called Shaivas (or Saivas) and are known to adhere to self-purification rituals as well as worship Shiv in a temple.
- Sam-Dam -the carrot and the stick policy
- Shradh - a ritual performed in memory of ancestors.
- Shastriy - classical
- Shikha - crest; a small lock of hair at the back of head. This lock is never cut.
- Shikha Vandanam - touching Shikha during Puja
- Shivam - auspicious.
- Shivling - a Shivling in general symbolizes the union of mind and soul.

- Shivsthan - place of Bhagwan Shiv.
- Shodh Granth - research treatise.
- Shramana - one who labours, toils, or exerts for some higher or religious purpose.
- Shudr - one of the four Varns.
- Sundaram - beautiful
- Sursa - a character in the Ramayan.
- Sut-Putr - son of a charioteer

## T

- Tap - tenacity, religious austerities or perseverance to make use of a source of energy, knowledge.
- Tapasvi - ascetic, hermit
- Tapshchrya - to lead an austere life.
- Tarpan - libation, i.e. offering of water, mixed with seed or grains, to a deity or ancestor.
- Tathagat - often thought to mean either one who has gone or one who has thus come or sometimes one who has thus not gone. It refers to Buddh.
- Tretayug - also called the silver age, lasts 1,296,000 years and the process of self-realization is the performance of opulent *Yagys* (sacrifices). The average life span is 10,000 years and the godly qualities decrease one fourth compared to the Satyug. It is during this age that Varn-Ashram Dharm is introduced.

## U

- Upwas- fasting for religious reasons.

## V

- Vaangmay - literature.
- Vaidya - versed in Ayurvedic medicine.
- Vaishnavism - is the worship and acceptance of Bhagwan Vishnu, “The Pervader” or “The Immanent” or one of his various incarnations (avtars) as the supreme manifestation of the divine.
- Vaishy - a member of the third of the four Varns, comprising the merchants and farmers.
- Vanvasi - forest dweller/ forester
- Vimarsh - discussion / consideration.
- Viraat - gigantic, enormous, great, broad.
- Vivah - marriage
- Vriti - is a technical term in yoga meant to indicate the contents of mental awareness
- Vyakhyakar - interpreter / translator

## Y

- Yagy - literally means sacrifice, devotion, worship, offering, and refers in Hinduism to any ritual done in front of a sacred fire, often with Mantrs.
- Yagypurush - the presiding deity of all *Yagys* i.e. Bhagwan Vishnu.
- Yaksh - is a broad class of nature-spirits, usually benevolent, connected with water, fertility, trees, forest, treasure and wilderness.
- Yugdharm - an aspect of Dharm that is valid for a particular Yug.

# Dr. Ravindra Shukl

Dr. Ravindra Shukl was born to Sh. Mannulal Shukl on 29 March 1953 in Village Raipur of Uttar Pradesh. His educational qualifications are L.L.B., M.A. (Hindi), D.Lit.

He was a member of Legislative Assembly, U.P., from Jhansi for four consecutive terms. He is a former Minister of State for Agriculture, Agriculture Education and Research, former Minister of State for Basic Education, Non-formal Education, Adult Education and Training.

His contribution to literature consists of *Sankalp*; Vande Bharat Mataram; Maa; Nagapati : Mera Vandan Le Lo; Shri Shatrughan Charitra (Epic). Prose works include *Varn Vayavstha: Maulik Avdharna Aur Vikriti*; Abhimat (under publication); Shiksha ki Pran Pratishtha (under publication). He has many published research papers to his credit and many students are researching on his literature.

Many literary awards and other honours have been bestowed upon him in the last twenty years.

He has widely travelled within India and in Sri Lanka.

Dr. Ravindra Shukl is an epitome of revolutionary spirit and is a staunch nationalist thinker. He keeps participating in various literary national and international seminars.

The grim situation caused by Covid-19 virus could not dent his spirit of social welfare to uplift poor and down-trodden and his literary interests. To promote Hindi, *Bhartiy* culture and civilization, he formed an organization 'Hindi Sahitya Bharati' (International). Under his stewardship as international President, 'Hindi Sahitya Bharati' has spread its wings in all *Bhartiy* states and thirty countries.

Contact Details
094150-30895, 0510-2443500
kaviravindrashukla@gmail.com
hindisahityabharati@gmail.com

Address
Sankalp 702/1, Civil Lines,
Jhansi - 284001
Uttar Pradesh

# Dr. Kumud R. Bansal

Dr. Kumud was born to a highly educated Shri Ramanand Bansal and a morally and spiritually strong Smt. Shobha Devi in 1956. The traditional landlord family of 1950's did not stop her from getting school education from Birla Balika Vidyapeeth, Pilani. She did graduation in Law, post graduation and Advance Diploma Course in German Language from Punjab University, Chandigarh, besides a short course in Portuguese language from Berlitz school of languages, Frankfurt, Germany and diploma in theosophy from The Theosophical Society (Chennai).

Her contribution to literature consists of *Sambandh Sindhu; Smriti Manjri; Jivan Kala; Todo Janjire; Lagi Lagan* (anthologies of Haiku). *Hey Vihangni; Jhanka Bhitar* (anthologies of Taanka). *Re Mann; Anubhut Lehre; Sapandan; Bargad ki Jaden; Jhina Ujas; Mann Banjara; Mann Jogiya* (7 anthologies in free verse). *Chintan* (a collection of essays). *Jadon ki Talash : Ek Sukhd Anubhav* (a book on family History). *Aanand Utsav* (a book on folk traditions). *Vibes* (An anthology of poems in English).

She has edited 27 books on various topics and 2 volumes of law journal and translated *Urgency of Change*, a book by J. Krishnamurti into Hindi.

Dr. Kumud is an avid traveler, who has extensively travelled in 28 countries. She is founder member of many NGO's and is associated with these in various capacities. She has been honoured with Excellence Award by National Commission for Women.

She is former Director of Haryana Sahitya Academy and Haryana Urdu Academy, Panchkula.

Contact Details
97290-46722; 88377-49835
forestriver.kumud@gmail.com

Address
Post Box No. 27,
Sirsa - 125055
Haryana

www.ingramcontent.com/pod-product-compliance
Lightning Source LLC
LaVergne TN
LVHW101939220826
846093LV00006B/60

* 9 7 8 9 3 5 4 8 6 6 0 4 3 *